W E C O U L D N O T F A I L

We Could Not Fail

*The First African Americans
in the Space Program*

RICHARD PAUL & STEVEN MOSS

University of Texas Press
AUSTIN

Requests for permission to reproduce material from this work should be sent to:
    Permissions
    University of Texas Press
    P.O. Box 7819
    Austin, TX 78713-7819
    http://utpress.utexas.edu/index.php/rp-form

♾ The paper used in this book meets the minimum requirements of
ANSI/NISO Z39.48-1992 (R1997) (Permanence of Paper).

LIBRARY OF CONGRESS CATALOGING-IN-PUBLICATION DATA

Paul, Richard, 1959–
    We could not fail : the first African Americans in the space program /
by Richard Paul and Steven Moss. — First edition.
        pages   cm
    Includes bibliographical references and index.
    ISBN 978-1-4773-1113-4 (paper : alk. paper)
1. United States. National Aeronautics and Space Administration—Officials and
employees—Biography   2. African American professional employees—Biography.
3. African American engineers—Biography.   4. African American astronauts—
Biography.   5. United States. National Aeronautics and Space Administration—
Officials and employees—History.   6. United States. National Aeronautics and
Space Administration—Rules and practice—History.   7. Discrimination in
employment—United States—History—20th century.   8. Race discrimination—
United States—History—20th century.   I. Moss, Steven, 1962–   II. Title.
III. Title: First African Americans in the Space Program.
TL521.312.P39   2015
629.4092′396073—dc23

                                                                2014030513

doi:10.7560/772496

For Renee

—RICHARD PAUL

With love to Lewis M., June, and Susan Moss

—STEVEN MOSS

Contents

Preface

In 2008, the alumni office at Southern University–Baton Rouge helped us find Frank Williams, one of the first six black engineers at the National Aeronautics and Space Administration (NASA). We contacted him at his home in New Orleans and asked if he would sit for an interview. "Yeah, sure," he said, though he quickly added, "But you should talk to my daughter," and handed her the phone. Tarsia Williams was suspicious. "What's this about?" she asked. We explained that we were producing a documentary and that her father was important to its story. There had been an article in the *New York Times* in 1964 that called him "a social pioneer." He was one of the first African American engineers at NASA. She was aghast. "He was what?" She never knew there was an article in the *New York Times*. She never knew he was one of the first black engineers at NASA. She knew none of this. Now she was excited, but she was still concerned. "My father is very sick," she said. Still, we arranged to conduct the interview, hiring a public radio producer in New Orleans to record Mr. Williams's half of the conversation.

When the day of the interview arrived, things did not go well. We asked the first question and all Frank Williams could do was cough—a long painful spasm of coughing that went on for nearly half a minute. A second attempt was no better, or a third. After about five minutes, Tarsia came back on the line. She was in tears. "I'm so sorry," she said. "I told you my father is very sick. He has mesothelioma, and I don't think you're going to be able to get him to do anything but this. And I'm so sad," she continued. "This was going to be my chance to hear all these stories. I'm sorry," she said. The interview was over.

Several weeks later, she sent an e-mail. Her father had died. At the funeral, they invited people to come and speak if they had a story to share. She said a man named Morgan Watson got up. She had never met this man before, but he had all kinds of stories about her father — about being with Frank at college; about how they went together to NASA in Huntsville, Alabama; about how Frank had a car and all the guys would jump in it on weekends for road trips to Nashville or over to Atlanta. About crashing the "whites-only" lunch counter at a late-night bus station, and how Frank Williams stood up for their rights. She had never heard any of these stories. When Morgan Watson was done, Tarsia went up to him — at her father's funeral — and said, "You know, there's a man making a documentary about the first African Americans at NASA. Can I give him your number?"

We Could Not Fail

Introduction

On May 13, 1961, in its first issue after Alan Shepard's historic Mercury mission, the *New York Amsterdam News* ran a front-page column by James Hicks that asked a question weighing on the minds of millions of Americans. "If you are like me," the executive editor of the nation's leading black newspaper said, "as soon as you finished thrilling to the flight of the United States's first man into outer space, your next thought was, 'I wonder if there were any Negroes who had anything to do with Commander Shepard's flight?'"[1] More than fifty years later, it is doubtful that many Americans could answer that question. It is safe to say that most know the name of the first black player in professional baseball and of the first person to integrate the University of Mississippi. Yet how many know the name of the first African American technical professional at Cape Canaveral, or of the man who integrated the Florida Institute of Technology? How many Americans know that the same man did both?

This book tells a story about a particular group of men who went to work either for the civilian or military space program during the period we now think of as the civil rights era, and the challenges they endured to accomplish what they did at the time and in the place that they did it. NASA tried desperately during this period to convince African Americans to move to Alabama, Florida, Mississippi, and Texas to work in the burgeoning space industry, but because of the South's well-deserved reputation for discrimination and violence, many would not go. Looking at the stories of those who did offers an opportunity to see an alternative to the standard civil rights narrative of marches, sit-ins, and lawsuits brought by the U.S. Justice Department or the National

Association for the Advancement of Colored People (NAACP). This is because during some of the most tumultuous years of the twentieth century, these men kept their noses firmly to the grindstone. While opposing forces, black and white, fought each other in the courtrooms and out in the streets over access to schools and public facilities, they kept their heads buried in their work. As "clashes between the increasingly militant Negroes and extremist whites created an atmosphere of crisis," they did not march, did not protest; they did not sue; they did not make threats.[2] Others did those things, sometimes right nearby, but these men instead realized a different kind of civil rights victory—quietly breaking through color barriers in education, employment, and politics to end up reviving and governing formerly defunct black towns, integrating southern colleges, earning PhDs and good jobs in advanced fields, and patenting important new inventions.

NASA's role in southern desegregation remains an unwritten and almost forgotten chapter in the history of the space program. There is much to say, however, about how the agency assisted portions of the South in stepping away from segregation as the Space Age promised to create a new society and shoot it off into the stars. The work NASA did as part of federal civil rights efforts, as well as the social consequences of its presence in the South, bridges two great American stories of the early 1960s. Technology and race are core issues in American history. From the cotton gin and slavery to the Space Age and civil rights they run together, and every once in a while they merge. This is the story of how these two principal themes of race and technology came together in the years before there was a Civil Rights Act, when civil rights laws and policies were just getting on their feet. During this time, many in the government were as committed to grounding Jim Crow as they were to landing a spacecraft on the Moon. This book tells that story in full and focuses on a group of brave and determined people who used the opening provided by this confluence to challenge a violent status quo.

Portrayals of the early space programs suggest they were an almost thoroughly white endeavor with no input from or impact on African Americans. NASA was German rocket scientists commanding legions of white American technicians, preparing and controlling capsules manned by jocular white fighter pilots and stoic white engineers. According to Konrad Dannenberg, a deputy to Wernher von Braun, the director of Marshall Space Flight Center (MSFC) in Alabama, "I am not aware of any high-level colored people who became part of the team."[3] Make no mistake, however. African Americans were there, and their presence had an impact.

While it is true that African Americans went to work elsewhere in the federal workforce, no other federal agency existed because of and for the Space Age, and no other component of the federal government enjoyed the romantic hyperbole associated with NASA in the early years of its human spaceflight program. The front pages of the black press portrayed African Americans working for the space program in the early 1960s as heroes.[4] That did not happen to people working at the General Services Administration or the post office. The image making rubbed off. As Morgan Watson, one of NASA's first black engineers, put it, "to be selected to participate at NASA was certainly a thing of pride."[5]

The men profiled in this book have stories that conform to significant trends and activities related to the space program's connections with the civil rights struggles of the 1960s. This is true even though their stories were unearthed through a completely random, journalistic process. It is more than coincidence that this happened, however. While these stories represent a narrow slice of the overall African American experience during this period, they are representative of a particular subset. The government engaged in a massive campaign to open federal workplaces and federal contractor workplaces to African Americans. Though the campaign did not come close to reaching the level of integration it sought, it still resulted in many successes. These men are among those successes. That is why their stories track so closely with the broader national narrative. Each chapter takes advantage of that relationship to tell the pioneers' stories as they orbit around national events, stories in NASA's southern host communities, and other narratives of importance to this largely unexplored confluence.

Julius Montgomery began building missile components at Cape Canaveral in the mid-1950s, a time when Florida undergirded and undercut its sunny, friendly reputation with a strain of savage Jim Crow segregation. Montgomery's story provides both the opportunity to discuss the broad impact of that segregation and a gripping demonstration of what it felt like on a person-to-person basis. His arrival corresponded with the climax and tragic end of the campaign of Florida's first civil rights martyr, Harry T. Moore, who registered hundreds of thousands of African American voters before the Ku Klux Klan murdered him only a few miles from Cape Canaveral. Moore's story, in turn, provides an opportunity to describe the Klan's ubiquity in the Cape Canaveral community and helps explain why Montgomery behaved the way he did when confronted by almost-daily racial assaults. Civil rights histories tend to talk about mass movements or to place individuals within mass movements. An unusual aspect of the stories of the first African Americans to integrate NASA and the space program is that they did not serve as a vanguard. Most acted as individuals and remained individuals, their individuality often constraining their behavior. So when a co-worker confronted Montgomery and declared, "You are nothing but a nigger," the result was behavior appropriate to the time and place. That sense of appropriateness is also evident in the story of Montgomery's integration of the Florida Institute of Technology (FIT).

While civil rights history is replete with stories of African Americans integrating southern colleges in the late 1950s and early 1960s, it is fair to say that the story of Julius Montgomery's integration of FIT is one of the least known. It happened without incident, but not without drama. It also had important ramifications. That it happened when it did—only a few weeks before the Greensboro sit-in—makes its conclusion significant. Montgomery's story is one of many in this book of people overcoming Jim Crow in the South using methods other than those we have come to expect. Also unconventional is the image of Florida that Montgomery's story presents.

The Florida story is also a significant part of the narrative of Theodis Ray. Ray grew up under unusual circumstances; he was the descendant of African Americans who fought for and won their freedom in the Civil War and who created a free black enclave whose destruction made way for the creation of Kennedy Space Center. The town, called Allenhurst, was itself a triumph. It was a place where African Americans worked for themselves rather than as vassals, and provided their children with education. The unique nature of life in Allenhurst molded Ray's self-definition, even after Space Age progress wiped out the place that he loved, leaving him and the rest of his community to scuffle for janitorial work at NASA. Ray's story also allows a window into the motivations behind the Kennedy administration's early attempts at workplace integration.

If there is a distinction to draw between a co-worker saying "you are nothing but a nigger" and a supervisor saying "you are qualified to be a senior member, but because you are so advanced for a Negro, we thought you were content," it is a distinction without a difference. Therefore, the story of Frank Crossley, while not laced with overt racism like that of Julius Montgomery, is equivalent. Crossley, who first heard and learned the true meaning of the term "equal opportunity employer" at precisely the time the Kennedy administration was compelling NASA contractors to advertise themselves using that term, has a story that allows an exploration of what the Kennedy administration did once it felt motivated to act for racial equality.

The literature of the civil rights movement has multiple tales of the U.S. Department of Justice and civil rights organizations threatening to cut off federal funding, staging sit-ins, holding boycotts and protests, and filing lawsuits.[6] Less well documented, however, is the role federal hiring and federal contracting played in the fight against racism and segregation. Jobs were the muscle the Kennedy administration had at hand when it came to forcing equal employment opportunity, especially in the South. Vice President Johnson, the head of the President's Committee on Equal Employment Opportunity (PCEEO), saw six agencies as leading the way.[7] NASA was one, and it was a large one.

Kennedy chose to rely on federal hiring and contracting because he doubted Congress would give him the power to do anything greater through legislation. Johnson, however, actually saw jobs as a vehicle to achieve racial integration. Once pressure at home and concerns about America's image abroad forced Kennedy into the civil rights struggle, NASA's location in the South became a problem and an opportunity. During his presidential campaign, Kennedy made the elimination of poverty a priority. Vice President Johnson had ideas on how to make that happen and to end racial tension at the same time.[8] They hinged in no small measure on the space program.

Johnson believed there was an inexorable link between southern poverty and southern racism. If an activist federal government could solve one, he thought, it could solve the other and transform the South. Even during the paternalistic days of the New Deal, the South largely held at bay federal money and the strings attached to it. By the 1960s, however, the death of cotton-based sharecropping had enabled the government to begin finally making the kinds of inroads that had until then not been possible. Johnson hoped to use this intervention to transition the South away from farming and toward technology, thereby bringing it into the nation's social and economic mainstream.[9] According to James Jennings, a retired NASA deputy administrator, it was common knowledge in African American communities that Johnson intended to use the space program to accomplish this goal. Once Johnson "found out that the Republican Party was interested in funding this program to beat the Russians to the Moon, he thought this was a good opportunity to have a federal program that they could get blacks into—to integrate that part of the government, and also influence the local community."[10] Johnson was not shy about promoting this idea, especially after he ascended to the presidency, where he declared NASA to be part of his "Great Society" initiative—its federal money transforming poor southern communities.[11] It was when Kennedy placed Johnson at the heads of both his National Space Council and the PCEEO that the vice president first found himself in a position to implement his plan. For many African Americans who went to work for NASA, the plan worked, and Frank Crossley's story offers a ground-level look at how the experiment played out.

Crossley's experience offers yet another distinctive look at an African American space worker achieving equality through unconventional means. In his case, new theories of workplace management were combined with a knack for invention, a superior intellect, and the attention of the black press. An influential columnist oversaw Crossley's early career and used him as a harbinger of the racial changes that would begin after World War II. Though Crossley did help to bring those changes about, the means he used were not what his early chronicler expected.

Crossley was not the only space worker lionized by the black press. Like much of America, black newspapers succumbed to the lure of the Space Age, but unlike their white counterparts, they also used the era's promise to make side-by-side comparisons with the reality of segregation. The stories of Otis King and Ed Dwight provide the chance to highlight both of these tendencies while also exploring the concept of the "Space Age"—what the phrase meant and how that meaning left it open to exploitation by those who wished to use it in the cause of racial integration. When King and his cohorts at Texas Southern University carried signs reading "Space Age Houston, Stone Age Schools," and when the black press said Dwight would be the first black man on the Moon, it reverberated in American society. Why and how it did are virtually unexamined subjects in the literature of both the space race and civil rights.

Dwight's saga received what can arguably be seen as too much attention while it was going on, overshadowing the achievements of African Americans who truly made a difference both to the space program and to racial integration. The eminent astronomer George Carruthers, who built the first observatory ever deployed on another celestial body, is one. Discussing his work and career provides an opportunity to examine the culture within a space program that said it wanted to integrate but could never bring itself to do it. The application and impact of the cultural norms behind that dynamic in the South in the early years of the space program is another area not explored at any length before.[12] The lives and experiences of ordinary space workers—scientists like Carruthers as well as other workers—are illuminating. It is important to be aware of the actions and accomplishments of the leading figures in the space program. However, it is equally important to know the places where NASA and the space contractors looked for African American engineers, scientists, and technicians and where they did not; how, where, and whether NASA enforced fair housing rules to integrate neighborhoods near space facilities; and how the Civil Rights Act overlaid the civilian space program's rendezvous with its communities.[13] Understanding these things sheds light on where and why Vice President Johnson's plan to change the South through technology jobs worked, and where and why it failed.

The traditional southern social order had its defenders, but the space centers represented a lot of money, a lot of jobs, and a tangible piece of Space Age glamour. An attack on NASA was an attack on prosperity; even more, it was an attack on a positive future. As Vice President Johnson pointed out in 1963, the MSFC's home of Huntsville, Alabama, was "one of the top communities in the Nation from the standpoint of Government employment."[14] Under the circumstances, southern leaders could do little but accept or ignore the agency's actions, and as a result, NASA affected long-held southern attitudes and actions on the issue of race, with varied success. In Brevard County,

Florida, money and Space Age symbolism had a mild impact on communities near Cape Canaveral. The Mississippi Test Facility in rural Mississippi and Houston's Manned Spacecraft Center, despite their radically different environments, shared NASA's failure to alter racial views significantly in their host communities. Agency efforts maintained a higher profile and enjoyed greater success in Huntsville than in any other host community. The stories of Richard Hall, Delano Hyder, Clyde Foster, Morgan Watson, and George Bourda, all of whom worked there, demonstrate the reasons why. Hall and Hyder evinced the community's racial blind spot, which went hand in hand with NASA's, while Foster, Watson, and Bourda help demonstrate the actions NASA finally took once it was pushed hard enough to respond.

At the MSFC, even at the time when Governor George Wallace was calling for "segregation now, segregation tomorrow, segregation forever," NASA made important strides on civil rights.[15] After a period of ignoring Wallace's provocations and the complaints of African American NASA workers about racist activity within the MSFC personnel office, the agency, its contractors, and two African American employees combined in 1963 to bring about change. Their most significant achievement was in building a cadre of local African American technicians and finding and recruiting African American engineers. The agency's quiet early steps became explosive in 1964 when NASA administrator James Webb made national news by appearing to suggest that MSFC would move out of Alabama if the state did not change its ways on race. Center director Wernher von Braun told chambers of commerce around the state, "Alabama's image is marred by civil rights incidents and statements" and urged businesspeople to be more open to integration.[16] Bourda and Watson, beneficiaries of the heightened recruiting program, fulfilled Vice President Johnson's dream by going to NASA for training that served them the rest of their lives. At the same time, the attention drawn to their arrival at the agency, in Watson's words, "helped change people's perception of black people in the South." The space program, he said, "opened the door," and in doing so, showed that African American professionals existed, and that they could work among the nation's technological elite.

Watson perhaps best characterizes the biggest contribution made by the people profiled in this book—these men often fought racism just by showing up. While America's historical memory recalls those who marched and credits them with changing the nation's racial situation, the African Americans of the space program demonstrate victory gained another way.[17] They worked in the system, they worked with the system; they studied hard; and when they went to work they did the very best they could. There is a reason why there were no mediocre First African Americans in sports. A journeyman first baseman would have helped make the racists' point—that blacks did not belong in the Major Leagues because they were not good enough. That Jackie Robinson hit nearly .300 and led the National League in stolen bases his first year demonstrated to the nation that he belonged in the big leagues. In the same way, the first African Americans in the space program had to be at their best at all times. With all that was riding on their presence there, they could not fail.

This book combines the fields of space history, African American history, southern history, and social history to tell its story. Bringing these fields together requires an analysis of a segregated society that embraced technology and promoted a federal agency that called for desegregation. It further requires an understanding of government policy and agency culture. There must also be an awareness of African American and white southern society. Finally, this is not just a story of great men and their deeds. It is the story of people relegated to the lowest rung of the social ladder in the Deep South who sought and built better lives for themselves.

We based the analysis contained in this book on the historical, economic, and literary works relevant to any discussion of the space program and the civil rights movement. However, this book also has a unique additional ingredient. Between 2007 and 2013, we tracked down many of the first generation of African Americans in the space program and placed them in front of microphones.

This process began when Richard Paul found a 1958 article from *Ebony Magazine* titled "Negroes Who Help Conquer Space: Over 1,000 Negroes Are in Satellite, Missile Field," and proceeded to track down the twenty men it profiled. All of those early missile men were long dead or, their relatives thought, too mentally incapacitated to be interviewed, but beginning with that article and leads from Steven Moss's thesis, "NASA and Racial Equality in the South, 1961–1968," and working through historically black college and university alumni offices, Social Security records, and phone call after phone call after phone call, we found many more of the first generation of African Americans who went to work for NASA when it opened in 1958. We found more by contacting prominent African American scientists and engineers, asking them for the names of African American professors they had while in college, contacting those professors, and asking them about colleagues and other contemporaries who worked for NASA or the space program.

The tales of their lives, their careers, NASA, Jim Crow, and how those elements fit together constitute the largest extant collection of oral histories ever conducted of African American employees of NASA and the space program. Their rich, untold stories and casual every-day observances offer an opportunity to look at the agency's action and the civil rights movement from the bottom up and thereby "gain a better sense of the less contentious ways in which minorities created advancement opportunities for themselves and in which the agency aided their advancement."[18] While archives, for example, offer NASA's perspective on why it could not hire blacks,[19] they do not yield vivid descriptions—like those offered in this book—of the terror that caused blacks to forego good jobs in NASA's southern communities.[20] The archives might illuminate what NASA administrator James Webb did to press his personnel directors and equal employment compliance officers to hire and promote more African Americans. Only oral history interviews, however, can explain how an African American technician broke the back of NASA's whites-only advanced training program in Alabama and the impact that action had on the personal and professional lives of black NASA employees.

The rockets that took astronauts into space did not emerge spontaneously from the hand of Wernher von Braun. People hammered and welded rocket parts into place; strung cables and wires; and in this era before computers, rigorously tested these vehicles with pencils and slide rules. African American space program workers were fully cognizant of the turmoil around them in the 1960s. Like their white colleagues, they watched it on TV. They also read black newspapers, listened to Martin Luther King Jr., and hoped for a world where respect and equality would replace segregation as the normal way of life, with higher wages and the right to use the same toilet as a white co-worker. NASA as an agency and its contractors as corporations did not cover themselves with glory when it came to hiring or promoting

African Americans, but the African Americans who went to work for both were able to assist in redefining racial identity in late twentieth-century America. James Hicks of the *New York Amsterdam News* wondered "if there were any Negroes who had anything to do" with significant elements of the space program. This book answers that question and explains why it matters.

A Man of Firsts

Julius Montgomery

I walked into this barracks full of all these guys — white guys — and I said "Good God!" I sat on my bunk and said, "How in the world will I be able to identify them? They all look alike to me."

JULIUS MONTGOMERY

Julius Montgomery's first day in the space program was lonely and terrifying. Walking down the dusty road past the squat wooden buildings at the entrance to Cape Canaveral Air Force Station, Montgomery was entering a place that would soon come to embody the very idea of Tomorrow in the American imagination. But what he faced that day in 1956 was a dispiriting combination of the sad and hateful present, tinged with the bitter history of yesterday. Montgomery was the first African American hired as anything other than a janitor at the Cape, but he shouldered a burden other racial pioneers did not. His experience was unlike that of Jackie Robinson when he integrated baseball, unlike that of the Little Rock Nine, who just weeks later would integrate Central High School in Little Rock, unlike that of Guy Bluford and Mae Jemison as they waved and boarded the space shuttle.[1] As he made his way to the building that housed the RCA Development Lab, there were no reporters along to watch, no columnist from the black press cheering and urging progress. There was no one from the National Urban League or the NAACP standing by to offer legal or moral support. Julius Montgomery was completely and utterly alone. Reaching the lab, he swung the door open and there faced a roomful of angry white men.

Sunshine Segregation

From the end of World War I through most of the twentieth century, Florida, where NASA would launch rockets to the Moon, was a terrible place to be an African American. In the 1920s, the state enjoyed a sustained land boom and cultivated its reputation as a vacationer's paradise. As it did, Florida—along with the rest of the nation—looked the other way when it came to horrific racial secrets such as the Rosewood Massacre, where whites burned and destroyed the black section of town after armed African Americans tried to defend their homes from a mob.[2] Florida had Jim Crow racial separation as severe as any other state in the former Confederacy—separation that was not just socially isolating but that also translated into deficits in government services that kept blacks running a race in which they could never catch up.[3] Most pernicious was the impact of discrimination on the public schools. Southern states spent one-third to one-half as much on education over the years 1890–1940 as the rest of the country, and that was for working-class whites.[4] For blacks, especially in the countryside, the gap was much more severe, and it was crushing.[5] In the area near Cape Canaveral in 1937, for example, the school board spent $69.05 per capita for white students and $27.04 per capita for blacks.[6] African American children were "crowded into very inadequate buildings and taught by poorly qualified teachers."[7] Sadly, that was actually an improvement from twenty years earlier, when Floridians elected Governor Sidney J. Catts, who ran on a platform opposing *any* education for blacks.[8]

Along with the deprivation came a capacious dose of terror. Because of a lack of legal protection (there were no black police and African Americans had not served on juries in the South since the 1870s), whole African American communities were under constant threat of violence and death.[9] It is a sad fact of the late nineteenth and much of the twentieth century in America, especially in the South, that African American lives were in the main considered worthless. That was not the case, of course, before the North won the Civil War. For more than two hundred years, African American life had had a price. Africans were property; bought and sold like a mule or a scythe. Black lives were not for sale after the war and to many that meant they now had little or no worth—a principal motivating factor within the sociogenesis of lynching.[10] Whites sometimes used lynching as a last resort, when no other form of coercion worked to keep the black population in line. Sometimes they just did it because they could.[11] This was especially true in the swampy mangroves surrounding Cape Canaveral.

Up through the time Julius Montgomery walked through that door, the Ku Klux Klan controlled East Central Florida. The sheriff of Orange County was a Klansman. There were city commissioners, aldermen, and county commissioners in the Klan. "Local businessmen joined the Klan almost like joining the Rotary club."[12] The Klan was so central to life there that the local paper covered their activities on the society page.[13] And in the Florida Klan's wake came the lynching. By the end of World War I, 95 percent of all lynching in the United States occurred in the states that formed the Confederacy,[14] and in the southern states where NASA was based (Florida, Alabama, Texas, Louisiana, and Mississippi), Florida had the highest lynching rate per capita.[15]

Montgomery knew all of this. He was from out of state, but Florida had a distinct reputation among African Americans. He knew about black men who looked at white men crossways and disappeared in the middle of the night. Consequently, he expected harassment from his co-workers. It would not surprise him if he got shoved, or if maybe someone spit. He knew quite possibly he could get punched. He faced his co-workers on that first day of his new job armed with nothing more than his knowledge of the rules of southern life, along with his wits and the paperwork that said he belonged there.

That he qualified for the job, which involved tracking, timing, radar, and telemetry, as well as repairing missiles, was a tribute to nearly twenty years of work by African American activists who had pushed the federal government into a grudging shift in policy on race relations. Not that many years before, when Montgomery's father was entering the workforce, say, an African American in the Old South with a degree in a field like mathematics or science could aspire to be a teacher and not much more.[16] One with a degree in engineering might build roads if the community needed roads and was a community that allowed African Americans to work in road construction.[17] If not, that person could become a teacher. Knowing this, Montgomery studied to be a linotype operator in college at Tuskegee Institute. He understood the paucity of jobs southern society would allow blacks to hold. Janitor and "concrete work" were the main ones, he said, "but you couldn't be the boss." Though he was interested in science, he had learned from a cousin's experience just how hard it would be to become a doctor. "My cousin had to go to Morehouse in Georgia, because they wouldn't let him go to the school in Alabama," he said. His cousin applied for medical school and they told him, "You can't go here, boy." Montgomery continued, "They would send you out of the state back in those days. And they paid for it! First, you have to apply to the school to be turned down by the state. And then they would give you an option of going to another school."[18] With employment options so narrow, he made a conscious choice to study a trade rather than to focus on academics. Then the air force drafted him and during his service he received his First Class Radiotelephone Operator's license. Despite that valuable asset, when he got out of the service and moved back home, he applied for lots of jobs but "got a letter back from them: 'We don't hire blacks.'"

In time, the people who worked at Cape Canaveral would be elevated in the popular imagination to the ranks of America's technical and intellectual elite. The black press would laud the African American ones with encomia. They were denizens of a "glittering new world"; part of a "team which sends US astronauts forth to master space."[19] None of that entered into Julius Montgomery's decision to work in the space program, however; he had a much more prosaic motivation. Montgomery was working at a black radio station in Mobile, Alabama. All he wanted was to find a job that would pay at least $100 a week. One day he got a telephone call from his mother. "'You got a telegram.' I said, 'Read it, Mama.' She said '$96 a week.' I said, 'Close enough!'" He gave two weeks' notice and headed for Florida.

As a southerner, Montgomery grew up under segregation. "I had not talked to a white person in my life until I was in the service," he said. "No conversation. That's the way it was" growing up. But because that segregation was complete, because he never encountered white people, he never had to feel their contempt. The government had desegregated the air force before it drafted him, so he had dealt with white people ("I remember walking into this barracks full of all these guys—white guys—and I said 'Good God!' I sat on my bunk and said, 'How in the world will I be able to identify them? They all look alike to me!'"), but the job at Cape Canaveral was going to be different. Would they accept him? Did they think he belonged? Did they even think— some of them—that he was human? "I was a strange person coming into an all-white building. All white." He entered the lab and eyed his co-workers. They stared back. "Nobody would shake my hand." His heart was pounding. Who knew the number of Florida Klansmen in the room? One by one as he approached these men, each one turned away. "I got to the last fellow," he said, "And I said 'Hello, I'm Julius Montgomery.' He said, 'Look, boy, that's no way to talk to a white man!'" At this point, there are any number of ways a man can react to these people he would be spending every day with for who knew how long. He could lash out; knock the man to the ground. He could go to management. He could storm out, call the NAACP, and file a complaint. But that was not who Montgomery was. This is who he was: he looked the white man in the eye and "I said—I said, 'Ah, forgive me, oh great, white bastard, what should I call you?' I really did say that! And I laughed, and he laughed and he shook my hand."

"Cease and Desist from Such Unfair Labor Practice"

The federal government had tried at least since the Franklin Roosevelt administration to address hiring discrimination within federal facilities like Cape Canaveral. Roosevelt imposed limited forms of affirmative action during the New Deal, and in fact the term "affirmative action" comes from the wording of the 1935 National Labor Relations Act.[20] The government integrated elements of the Civilian Conservation Corps in the 1930s and the Agriculture Department's Soil Conservation Service hired African American technicians (though, tellingly, regulations restricted that second group to only counseling African American farmers).[21] The most significant action taken by Roosevelt came in early 1941 when, hoping to head off a threatened African American march on Washington for equal rights, the president created the Fair Employment Practices Committee, which sought to end discrimination based on race, creed, color, or national origin in federal employment and defense contracts.[22] A second committee was established by executive order in 1943 (after the first committee collapsed) and its jurisdiction was extended to additional industries and to labor organizations, which it said had the "duty" to eliminate discrimination not only in employment but also in union membership.[23] Oversight by this committee was never strong and contractors often threw out their good intentions on racial integration in the rush to complete federal projects.[24] It nonetheless made some impressive inroads at key agencies that laid the groundwork for future progress.[25] A southern filibuster finally killed the committee in 1946.[26] The Truman administration created its own committee to oversee federal contractors and federal hiring. Eisenhower dissolved that and replaced it with yet another committee, this one headed by Vice President Nixon, which had taken several actions by the time Julius Montgomery went to work at the Cape.

While the government designed these programs to get African Americans in the door, their success often depended on the ability of people like Julius Montgomery to find ways to overcome the hatred and discriminatory attitudes and behaviors that kept them out of the workforce in the first place. Montgomery's way worked for him. Repeatedly—as he did that first day—he would size up a situation, and then defuse it. Disarm his opponent with a joke, preferably an audacious one. Over the years, he took plenty of opportunities to tell his coworkers, "Look, I'm part of the educational program to train you guys to act like people. You've been acting like rednecks all your lives. So you need training; retraining."[27] Was his approach the best way? There is a counterstory from the same place at about the same time that shows that it just might have been.

The Martyr of Mims, Harry T. Moore

The counterstory is that of a pioneering activist for civil rights who raised his following and fought and died in view (or certainly within earshot) of Cape Canaveral.[28] Today, Harry T. Moore of Mims, Florida, is a mostly forgotten early hero of civil rights. But his impact was enormous and makes clear the scope of the task set before those looking to use government programs like NASA to change the southern way of life. Moore was not a contemporary of the civil rights pantheon; he labored at a time when the NAACP battled only furtively in the South, when Rosa Parks was a newlywed and Martin Luther King Jr. was still going by the name "Mike."[29] Harry T. Moore was born in 1905 in North Florida and raised by his aunts in a home that infused him with an appreciation for education (his Aunt Masie had a PhD from Syracuse University), a sense of justice, and a strong motivation to fight for it. In the 1920s, Moore moved to Brevard County and took a teaching job at a black elementary school in Cocoa.

In his spare time, Moore courted danger by creating the Brevard County chapter of the NAACP. Already "known and despised by white officials all over the state" and fired for trying to organize his black colleagues, Moore turned his attention in the late 1930s to the two most pressing issues facing African Americans in the South.[30] Operating with a meager stipend from the NAACP, he threw himself into the battle over voting rights and the drive to stem the "window to the soul of white supremacy and African-American life in the South," lynching.[31]

The case that first spurred Moore to action was one where "a young colored boy sent a Christmas card to a white girl, who showed the card to her father. A posse of white men captured the boy, hogtied him, and forced the boy's father to watch as they tortured the boy and drowned him in the river."[32] While this kind of thing was sadly all too common in Brevard County, Moore was never a man to let an injustice pass, and according to William Gary, who has held the job Moore once had at the North Brevard NAACP, the community would not have let him anyway. "Speaking from personal experience," Gary said, "you get drawn into things that you could do without, because even today people are looking for someone to provide solutions to the many problems we face here."[33] After campaigning to draw attention to that first case, Moore began traveling across Florida, a self-funded investigator of and crusader against lynching. Taking on the white school authorities could get Moore fired, but taking on the Florida sheriffs, he knew, could get him killed.

His fight against lynching was not the end of it, however. Back when he was a teacher, the issue of voting rights had consumed Moore. This was a time when Florida still imposed a two-dollar poll tax that disenfranchised poor blacks and whites. A black voter with the money and guts to cast a ballot had to vote Republican. Even after the U.S. Supreme Court outlawed whites-only primaries, Moore was still battling as late as 1947 against legislation to make the Florida Democratic Party a private club. Some would have called Moore's effort an exercise in futility. Nonetheless, by 1950, a full fifteen years before the passage of the Voting Rights Act, he worked to make 116,145 African Americans eligible to go to the polls in Florida. That was 31 percent

of all voting-age blacks in the state, 98 percent of the blacks in Brevard County, and a rate 50 percent higher than anywhere else in the
South.[34] One hundred thousand people could be a potent voting bloc
in any statewide election; but these hundred thousand were not just
registered. They exercised discipline and organization—seized, as they
were, with the recognition that they could have an impact on their own
political interests.[35] By bringing these people together, Harry T. Moore
had caused a significant shift in Florida politics by the late 1940s.
White politicians—even those who ignored or openly attacked the
black community—now had to listen to their African American con-
stituents.[36] Those hundred thousand voters were poised to change the
1952 governor's race, though Harry Moore would not live to see it.[37]

A few days before Christmas 1951, he was told "he ought to have
his neck broke for putting notions in Niggers' heads."[38] Moore went
home to Brevard County to celebrate Christmas with his wife Harriette
and one of their daughters. That night was also Harry and Harriette's
twenty-fifth wedding anniversary. As the family celebrated, white
Klansmen slunk through the darkness to deliver their ghastly present.
They placed it at the base of the Moores' home, and at 10:00 p.m.
Christmas Eve, an enormous explosion rocked the town of Mims.
People who heard the bomb (and they heard it as far as Titusville, four
miles away) told the FBI they thought it was loud enough to be all
kinds of things—a tanker truck that had blown up or a propane tank.
There was another common thought that night, too. A rocket launch
was anything but a routine thing in 1951. FBI agents later reported
that one man ran into his yard thinking that a missile had blown up at
the Cape. The next morning, Christmas day, much of the black community made the trek out to the house, where they learned Harry
Moore was dead. "He died on the way to a hospital," Gary said. It was
not their local hospital, which would not admit him because he was
black. Moore's wife Harriette died nine days later.

For a short time, there were protests around the country. A group
in New York started a boycott of Florida citrus. The governor read their
letter "and said, 'This stuff is getting serious.'"[39] State authorities were
alarmed, not about the state's racial problems, but about the tourism
industry and the state's reputation. With considerable business interests at stake, they decided to end the growing uproar. Within days,
the Moore story "just vanished" from the *Titusville Advocate*. Within a

week as far as anyone could tell it was "business as usual" in East Central Florida.[40]

Florida could do that. The state has a unique and distinct mythos. In the 1950s, it was a place of sunshine and beaches where people spent all their time by the pool or going to look at the alligators and the Weeki Wachee Water Ballet.[41] Racism and racial strife in the 1960s bring to mind Alabama or Mississippi, not Florida. The violent death of Harry T. Moore showed otherwise. So did the day-to-day life of Julius Montgomery. The racism that confronted him on his first day did not let up as time went on. During all his years at the Cape, Montgomery said no one ever invited him or any of the other black employees when there was an office party or a going-away lunch. "We could not go, so we just didn't go. We were just not invited. That's just what it was."[42] In addition to the casual racism, there was also the overt. One day, Montgomery said, a co-worker walked up to him, looked him in the eye, and said, "You are nothing but a nigger." As he always did in these kinds of situations, Montgomery said he closed his eyes to compose himself. When he opened them, he said, "I looked down, the guy's on the floor." Montgomery said he never touched the man. "I swear to this day, I did not move." But there he was on the floor. Montgomery said he looked down, thought, "what are you doing down there? What in the world!?" and finally concluded, "God did it."[43] That was how he survived in Jim Crow Florida—self-composure, a sense of humor, and sometimes a little help from above. He walked away from the man and went back to work.

How We Got Here

Julius Montgomery's work did not involve astronauts and rockets. He worked on the military side of the space program. In fact, much of what he did was top secret. That sounds glamorous in the abstract, but aside from occasional international travel, it was mostly mundane. Montgomery was part of team called the Range Rats, who repaired malfunctioning ballistic missiles and the systems that tracked them. Whenever a missile blew up, the air force sent in the Range Rats to figure out why it happened and then fix it. Their work mostly involved developing circuits (there was no "over-the-counter equipment to do the jobs," Montgomery said) and soldering connections. The team was also sent out to the ships that "searched the skies for anything the Russians were doing" to perform maintenance on the satellite equipment.[44] That kind of work was consistent with the Eisenhower administration's orientation toward space: the space program was a bulwark against Soviet aggression. But government policy on that subject changed radically after the next presidential election. When it did, it had wide-reaching ramifications for the South and the drive for civil rights.

Anything viewed from the perspective of decades can seem inevitable, but the Kennedy administration's decision to put a man on the Moon was not. It was the culmination of a years-long offensive to drive government policy. In his most eloquent speech about outer space, President Kennedy asked a series of pointed questions that addressed the skepticism many expressed about what would be an endeavor of massive cost. "Why, some say, the Moon?" the president asked. "Why choose this as our goal? And they may well ask why climb the highest mountain? Why, thirty-five years ago, fly the Atlantic?"[45] Kennedy's answer to those questions is now famous. "We choose to go to the Moon in this decade and do the other things, not because they are easy, but because they are hard." That is, of course, not the reason why the United States spent billions of dollars in outer space. The real reasons were much more practical and strategic—the need to know what our enemies were up to and the need to stay ahead in an existential battle for nuclear superiority and international popularity.

The argument that ended with Kennedy's decision had begun fifteen years earlier, at the end of the Second World War. Science and technology brought horror and devastation during the war, but they also brought the conflict to its end and ushered in life-altering changes like jet engines, radio astronomy, radar, satellites, and missiles. The United States had come through the war physically untouched, and when it was over, American leaders—both policymakers and leaders in business, industry, and higher education—were able to pick up whichever of these advances they wished, and do with them whatever they wanted.

The war had also changed much about why and for whom American scientists did the work that they did. We can see perhaps the strongest evidence of that change in a lament expressed by the president who was in office while most of it was settling into place. In the annals of presidential farewell addresses there are—at best—two lines recalled today by the public at large: George Washington's warning against "foreign entanglements" and a line from Dwight Eisenhower's address, spoken as he poured out his frustration over forces he said were hijacking the federal purse. "We must guard," the president told his television audience that night, "against the acquisition of unwarranted influence, whether sought or unsought, by the military/industrial complex."[46] In this speech, the president complained about the direction the country was taking when it came to spending money for science and technology—a direction that started during the war. Of science he said, "A steadily increasing share is conducted for, by, or at the direction of the federal government." Politicians and the defense industry, he said, removed university research from its lofty perch for the benefit of the war effort. When the war ended, however, they declined to put it back. The result, Eisenhower said, was that universities had "experienced a revolution in the conduct of research. Partly because of the huge costs involved, a government contract becomes virtually a substitute for intellectual curiosity."[47]

At the time he gave this speech, Eisenhower's vice president had just been defeated in his run for the presidency. In important ways, this speech was really about that political transition. The speech was not a warning against defense contractors; instead, it was a bitter response to what had happened to the Republicans.[48] Over the previous years, from 1957 until the election, the Democrats had pounded Eisenhower, particularly when it came to spending on outer space. That carping, he thought, was a significant factor driving the Republicans and the Republican agenda from power.

Understanding the conditions that put America in a position to place a man on Moon requires a full recollection of the nation's state of mind during the Eisenhower years—in particular, to America's evolving relationship with the concept of nuclear destruction. At first, Americans had largely embraced the atomic bomb. Eighty-five percent supported the attacks on Hiroshima and Nagasaki in 1945.[49] In 1951, 51 percent of Americans were urging the use of the atomic bomb in Korea.[50] But as the decade wore on, a creeping dread highjacked the American psyche. In 1952 a nuclear test code-named "Mike" carved out a mile-long hole in the bottom of the Pacific Ocean and impressed on Americans the ominous power of the hydrogen bomb.[51] What made matters worse, from a U.S. perspective, was when these technologies, which had once been ours alone, found their way into enemy hands. In 1945, the United States exploded the atomic bomb, demonstrating our technological virtuosity. In 1949, the Soviet Union did the same thing. In 1952, the United States exploded a hydrogen bomb. In 1953, the Soviet Union did the same thing.[52]

Six years later, the Soviet Union launched *Sputnik*, the first human satellite. That was followed shortly afterward not by an American satellite launch (that attempt failed miserably), but by the launch of a political strategy in the United States to make Americans scared enough to care. While the military rocketry establishment was concerned about the *Sputnik* launch on May 15, 1957, the public at large and the president were not. Anthropologist Margaret Mead was conducting a field study of high school and college students at the time, and on the night *Sputnik* launched, she sent a telegram to her researchers asking them to add questions about *Sputnik* to find out what people were thinking and feeling; they found that the launch actually engendered a mix of concern, hope, and excitement.[53] President Eisenhower demonstrated no surprise or concern at the launch of *Sputnik*. The day it hit the press, Eisenhower left for Camp David to play golf.[54] When he got back, the president answered some *Sputnik* questions at a press conference, but at his next press conference, no one asked about it. That nonchalant attitude, however, would not last.

The strategy to get America concerned and excited about *Sputnik* evolved through typical Washington back channels. Thirteen days after the launch, a longtime aide to Senator Lyndon Johnson named George Reedy got a visit from his friend Charles Brewton. Brewton was an assistant to Alabama senator Lister Hill, who, like Johnson, saw federal spending and economic development as a way out of the South's poverty and isolation. Up until then, Reedy said he and Johnson had "put the space thing on the back burner." But Brewton had an idea that would change their minds. He gave Reedy a memo in which he said that exploiting *Sputnik* "could clobber the Republicans" and elect Lyndon Johnson as president. Reedy sat up most of the night composing a memo to Johnson on how to exploit the *Sputnik* issue, telling his boss, "This could be one of the great dividing lines in American and world history, the whole history of humanity."[55]

Republicans and Democrats had been battling over the best way to address the Soviet threat. Eisenhower preferred nuclear deterrence—mutually assured destruction. He believed, he said in a 1957 speech, that America had "enough power in its strategic, retaliatory forces to bring near annihilation to the war-making capabilities of any country."[56] Democrats leaned on a strategy called "flexible response" that called for a range of options. They argued that by cutting defense expenditures—relying more on nuclear deterrence and less on conventional forces—Eisenhower had made the United States more vulnerable.[57]

Reedy told Johnson that the Democrat's key to success was to pound away at *Sputnik* as evidence that America had fallen behind the Russians and to support spending on a civilian space program. The idea fit well with a key desire of Johnson's. He had been looking for a way to pump massive amounts of federal money into the South to try to change its economy and—he thought—its racial dynamic, too. Space, Reedy told him, was the key he had been looking for.[58] It could change the South permanently from an agricultural region to one based on knowledge and technology; it could simultaneously bring glory and honor to the United States and (reflectively) to Lyndon Johnson.

Johnson picked up the ball and ran with it. "In the Air Age," he said in a speech shortly after the *Sputnik* launch, "we were powerful because we had airplanes. And now, tonight, the communists have established a foothold in outer space."[59] Insinuating that Eisenhower's inattention had allowed the Soviets to catch up and surpass the United States, Johnson said, "It took the Soviets four years to catch up with the Atomic Bomb. It took the Soviets nine months to catch up with the Hydrogen Bomb. And now, tonight we're trying to calculate how long it will take us to catch up with the Soviet satellite."[60] President Eisenhower continued to believe that a space race with the Soviets would be expensive and counterproductive.[61] Johnson mocked that position, saying, "It is not very reassuring to be told that next year we'll put a better satellite into the air. Perhaps it will even have chrome trim, automatic windshield wipers."[62]

Contrary to what Johnson was saying, throughout the 1950s, the U.S. had been involved in extensive defense-related activity high above the atmosphere, had spent billions on rockets and missiles, and planned to spend more. "I don't think we should pay one cent for defense more than we have to," Eisenhower said at the time, "but [what]

I do say is our defense is not only strong, it is awesome, and it is re-spected elsewhere."[63] The president held steady to that position, but world events were about to push things over to Johnson's advantage. On November 3, 1957, the Soviet Union launched *Sputnik 2*. This second satellite launched with a dog and a payload that, had it been a war-head, was capable of doing considerable damage to the United States.[64]

Johnson was a member of the Senate Armed Services Committee. In the wake of the second *Sputnik* launch, he shared Brewton and Reedy's ideas with the committee's chairman, Senator Richard Russell, who authorized Johnson to use the Senate Preparedness Subcommit-tee to hold some hearings into the *Sputnik* question.[65] The subcommit-tee had been defunct for several years, but Russell had Johnson revive it to use as a cudgel to beat the Republicans. As the hearings opened, Johnson compared the *Sputnik 2* launch to Pearl Harbor, saying, "In some respects I think that it is an even greater challenge. In my opin-ion," Johnson said, "we do not have as much time as we had after Pearl Harbor."[66] As expected, Johnson's subcommittee found U.S. space ac-tivities underfunded and poorly organized.

As often happens in Washington, interest groups coalesced around this argument and came together over Johnson's hearings. Called to testify were generals and admirals; scientists like the man known as "the father of the hydrogen bomb," Edward Teller; academics like Van-nevar Bush, dean of the MIT School of Engineering; and aeronauti-cal leaders like Wernher von Braun, the German rocket scientist and future director of NASA's Marshall Space Flight Center, and his com-mander at the Army Ballistic Missile Agency, General John Medaris.[67] Their motivations varied, from those who saw spaceflight as man-kind's destiny to those in the aerospace industry who viewed increased spending as an opportunity; to university-based scientists who had been receiving millions in federal contracts during the Second World War that were now winding down. Those who did not testify wrote op-eds and magazine articles and made television appearances warn-ing of the dangers posed by Soviet control of space. They advocated for massive spending on manned spaceflight to demonstrate America's technological and scientific prowess in an effort to ward off any fear at home or abroad that the nation was second best.[68]

This line of thinking tracked along with that of liberal Democrats. In the late 1950s, they were arguing that Eisenhower's focus on balancing the budget and returning money to the private sector had taken funding away from research and development efforts to create valuable things that could benefit the nation—like new ballistic missiles and spaceships. Liberals said Eisenhower was instead making that money available for R&D on consumer and luxury goods.[69] They pressured the White House both to demonstrate a devotion to the future (and to the fight to beat the Russians) and to institutionalize the space race by creating a federal Department of Space or a cabinet-level Secretariat of Science.[70] The pressure worked, forcing Eisenhower to make large and important structural changes to the United States government— changes not matched in scope until the creation of the Department of Homeland Security after the terrorist attacks of September 11, 2001.[71] First, he created the Office of Presidential Science Advisor, the first person in the White House whose sole purpose was to speak for science and technology. Second, he created the Advanced Research Project Agency, which later became DARPA, the agency that created the Internet. And finally, he streamlined the outer space components of the Defense Department, largely by creating the most visible legacy of the *Sputnik* launch, NASA. Eisenhower signed the National Aeronautics and Space Act of 1958 on July 29, and the space agency began operations on October 1, 1958.[72]

These changes are what Eisenhower was complaining about in his "military/industrial complex" speech. During the previous three years he is said to have held to the position that "Lyndon Johnson can keep his head in the stars if he wants. I'm going to keep my feet on the ground."[73] But it had not worked. Eisenhower was furious that the liberals had forced him to spend money in places that did not need it. That spending was about to launch into the stratosphere, he thought, and he intended to let the American people know what they had just allowed to happen. The government, he said, needed to "maintain balance in and among national programs—balance between the private and the public economy, balance between cost and hoped for advantage—balance between the clearly necessary and the comfortably desirable."[74] Johnson had another agenda and, assisted by circumstances, was able to bring along the president he entered the White House with to see the benefit of space spending as a tool for social change.

Although the National Aeronautics and Space Act of 1958 never charged the space agency to be an agent of social change, reforms did happen because of NASA's presence.[75] NASA's challenge to racism in its southern host communities had several components: NASA centers would obey and enforce presidential executive orders, laws passed by Congress, and directives issued by NASA headquarters, applying economic pressure at local and state levels, as intended by the federal government. In so doing, they would help to convince business and civic leaders to pursue desegregation. But more importantly, NASA officials were able to use the agency's enormous prestige and public image as political leverage. This is not to suggest that this was deliberate; it was not. The government was not going to spend all of those billions of dollars just to make a point. The mission of NASA was space, but because it was in the South, it was going to make a difference in society as a whole.

First at FIT

Back on the ground in Florida, just about the time the Kennedy/ Nixon election was gearing up, Julius Montgomery was having yet another experience that demonstrated just how badly change was needed. Any building you go into, it seems there is a plaque on the wall commemorating someone—usually people who have done great things. Institutions honor them to try to keep alive the memory of their accomplishment. At the Florida Institute of Technology (FIT) they have an honor like this, though it is not a plaque on the wall. Every year on Martin Luther King Jr.'s birthday they give the Julius Montgomery Pioneer Award to someone who has made a significant contribution to the community.

Montgomery is not a benefactor; he did not give the school millions of dollars to get started. What Julius Montgomery gave FIT is more important than money. At a pivotal time in the school and the country's history, Montgomery once again put his pride aside, allowing FIT to remain segregated at harm to himself but to the benefit of the institution. His sacrifice and generosity were so significant that according to FIT's founder, "The school would have been tossed out on the street" and closed down if not for Montgomery.[76] The story of why FIT gives this award—like the rest of Julius Montgomery's story—reveals so much about the times in the early days of the Space Age. It also talks powerfully about the human spirit, the desire to advance, and the benefits of going against the grain of what we consider traditional civil rights protest.

FIT started as a night school, set up to meet "the educational needs of America's missilemen."[77] Companies staffing Cape Canaveral in the 1950s feared that the humdrum life of pre-NASA Eastern Florida was going to make it difficult to recruit and maintain the best technical people. So when Jerry Keuper, a senior engineer in the Missile Test Project, went to his bosses at RCA and told them he wanted to open a school to keep workers engaged and up to date on the state of the art in engineering, they gave the project their full support. The school's original name was Brevard Engineering College, and it began life in three rented classrooms at Eau Gallie Junior High School, a public school near the current site of the Melbourne, Florida, international airport.[78]

Not too long after opening, though, BEC "found itself in the midst of a local controversy"—a controversy started inadvertently by Julius Montgomery.[79] As Montgomery recalled, Dr. Keuper put up a notice on an RCA bulletin board asking people to sign up if they wanted to enroll at his new school. It also asked them to list where they had done their undergraduate work. "I signed it," Montgomery said, and "of course I put down Tuskegee."[80] Within days, the Brevard County superintendent of schools, Woodrow Darden, was on the phone to Dr. Keuper, telling him the school system was canceling BEC's contract to rent the classrooms. This same Woodrow Darden would tell the National Academy of Arts and Sciences, "The space program, as such, opens up new opportunities for the schools . . . It enables us to do more experimentation." That experimentation apparently included neither school desegregation nor integration.[81] According to FIT's official history, Darden told Keuper he was worried about traffic jams "and 'other things.' High among the list of 'other things' that made BEC a problem for Woody Darden was the school's admissions policy."[82] Julius Montgomery said that Darden phoned Keuper and "he was told, 'I'm sorry, but this fellow here is from Tuskegee so he must be black. He cannot come to this classroom.'"[83] The ultimatum was clear: expel this black man or close the school.[84]

What happened next, Montgomery said, was that while he was at his workbench soldering some missile parts a message came over the intercom: "'Julius Montgomery, come to the office please.' So I went to the office. And the manager said, 'I want you to meet Dr. Keuper. He wants to talk to you.' I said, 'Yes, Dr. Keuper, what can I do for you?' He said 'Well, I need your help.' And the help they needed was for me to take my name off of that list so that he could start his new college." It would only be for a short time, Keuper promised. As Montgomery remembered, Keuper said, "As soon as I get my own buildings, you are welcome."[85] It was 1959—five years after *Brown v. Board of Education*, four years after Rosa Parks stayed in her seat and said "no," three years after a bus boycott in Tallahassee, and just a few months before young men would sit down at a segregated lunch counter in Greensboro, North Carolina. Now, faced with the same choice, what would Julius Montgomery do? Support the system or fight it? Help himself or help his race? Montgomery said he sat and thought for a minute. Then, "I said, 'well, OK.' After a short conversation—for the better good of everybody—I took my name off the list." The school, he told Keuper, could stay segregated. He would drop out and he would not cause a fuss.

That the Florida Institute of Technology created an award in Julius Montgomery's honor speaks to the existential nature of his decision. Today FIT has 130 acres and nearly three thousand students. *U.S. News and World Report* ranks it among the top two hundred colleges in America.[86] But in 1959, its future was fully dependent on the decision of one man. The decision he made was a selfless one. With Montgomery out of the way, "Darden dropped his threat of immediate eviction." BEC could use the classrooms.[87] Several months later, after BEC got its own buildings, Keuper kept his word. The school desegregated and Julius Montgomery enrolled.

Some would see Montgomery's decision as redolent of Booker T. Washington's "Atlanta Compromise" speech, a sell-out to segregation. This was, by all accounts, a snap decision, however, and one individual made it. Of course, James Meredith was one individual when he integrated the University of Mississippi. Meredith was part of a movement, however, with the backing of the NAACP. In this situation, Montgomery did not see himself fighting the institutions of segregation or racism; he saw himself as helping a guy out and helping out co-workers who would benefit from attending this new school.

Montgomery's is one of many stories of the space program's African American pioneers that does not comport with the stories of civil rights achievement that are found in the standard high school textbook. Not everyone achieved the ends of the civil rights movement by marching, picketing, and saying "no." Many did, of course, but others, like Julius Montgomery, applied the principles of self-help and—often—accommodation to reach the same ends. It worked for him, but he admitted that the black community did not always understand. "I got a lot of problems from the black community," he said.

People never understood how it was that he got his job at Cape Canaveral, for example. "I had to explain to them that I took a test, you know; they gave me a test and I passed." He said people in the community did not understand his work ethic. "They could not understand how I could come down, go to work, on a regular job—they could not understand that," he said. Other things he did caused him problems, as well. "I read the paper every day and they couldn't understand that," he said. "They said I was acting white." The life he portrayed sounds at times like one of supreme loneliness. Along with the abuse that he endured all those years from whites, he got no credit as a pioneer in the black community. "They would leave me sometimes at work and I would have to hitchhike home," he said with a mordant chuckle.

The community turning its back on the civil rights hero and leaving him to walk home alone is yet another story that does not conform to what we read in school. Montgomery's story clearly stands out. It is a lesson perhaps that the civil rights movement, like so many other undertakings in history, is not monolithic. All mass movements—revolutions, wars, western expansion, the civil rights movement—begin, end, and gain their sustenance from individual choices.

There is solace, however, in knowing that despite the scorn from both sides, Montgomery was content to continue being a pioneer. He began standing for election to the city council in Melbourne in 1956, a time when fewer than fifty African Americans held elected office in the eleven former Confederate states and in a county where some old-line council members still had fond memories of the Confederacy.[88] Nevertheless, he pressed on and, as if being first at Cape Canaveral and first at FIT were not enough, he attained yet another milestone in 1969 when he became the first African American ever to win a seat on the Melbourne City Council.[89]

A story from Montgomery's first election is worth recalling as we turn from the 1950s to the 1960s. "First time that I ran for City Council in 1956, we had fourteen precincts," he said. "I took all of the precincts except one—except mine"—the black precinct. While he understood the reason for that, nonetheless, "I was really disheartened. I was really shocked." In time, he found an advisor, a white man from New Jersey who taught him an important lesson. "He said, 'Do you want to win, or do you want to be loved in your black precinct, which only had about nine hundred or six hundred votes? You can't win. You've got to go to the other side to get them.' So we did." Race matters in Florida, the advisor told him, but times are changing.

Change would come, albeit in an excruciating creep. In the coming years, a new presidential administration would flex new types of muscle—muscle that included the space program—to try to achieve racial parity in the workplace; at the same time, new styles of civil rights protest would emerge. The advisor's admonition was significant. You are a person of substance and a man of historical import. Understand who you are, he told Montgomery; do not dwell on who you think you are or who you think you should be. Know who you are and move forward.

"There Was a Lot of History There"

Theodis Ray

We just understood what we were and our identity and to not try to venture beyond that. We knew when to say, what to say, how to say it, and who to say it to.

THEODIS RAY

As Julius Montgomery learned in his political career, there are times in life when who you are is less important than how you define yourself. That was also certainly the case for Theodis Ray. His career in the space program was nothing out of the ordinary; his life was no more or less eventful than anyone else's was. What set him apart was his history, or more importantly, his sense of his history and that of his descendants. The conclusions he drew from that history defined him and configured his destiny—dictating how he navigated the difficult challenges of the society into which he was born. Ray was a native of Brevard County who grew up on the very ground that became Kennedy Space Center. Those were facts that he never let slip from his consciousness. His entire life, he was deeply cognizant of the proud history of the African Americans who grew up in that area, and more importantly, how they differed from others. "Most of [the other black people in Florida] came from plantations or places that the people owned where they stayed," Ray liked to remind anyone who would listen. Other blacks "lived in, like, a controlled basis, versus where we were."[1] People where he lived—people like Ray—were different. People there were free.

Whites may not have considered free blacks as "constituent members of our society," but in African American communities, especially after the Civil War, those who had never known enslavement constructed and then maintained a rarefied status for themselves. Whites saw slaves as property and lesser humans, so the desire to create a distinction between "bona fide" free blacks and those they called "sot [set] free" is understandable. It was not too long after the advent of universal emancipation that free black people began thinking of themselves as "colored"; neither black nor white, they established schools, social clubs, societies, and churches that they kept apart from the black communities at large, handing down a sense of separateness and an accompanying sense of superiority to their offspring.[2]

Coping with Yesterday in the Land of Tomorrow

Societal constructs around color were relevant when NASA set up operations in the early 1960s. The openings provided first by White House executive orders that tried to force equality in federal hiring and then by the Civil Rights Act created an irony that human resources personnel struggled with in the South. "White collar jobs were now to be available to black men," they knew, "but where were the black men to come from?"[3] Someone needed to step up to get out the word that these new jobs were available for blacks and then to train and hire them. There was a belief within NASA that "the most logical employer to initiate the kind of programs necessary to achieve true equal employment opportunity was the federal government; and NASA, being an arm of the federal government, and a relatively new one at that, afforded an industry-wide, nationwide opportunity for immediate equal employment." The space agency "was a natural," many of its people thought.[4] Well, yes and no, because no matter how fresh, new, and futuristic NASA's image was, reality dictated that it would operate within the restrictive parameters of the communities it called home.

During the Second World War, when the federal government set up the Manhattan Project, which created the atomic bomb, it built a number of communities entirely from scratch. But problems that cropped up in Washington State and Tennessee pushed the original NASA planners to take a different route as they contemplated the massive infrastructure they would construct to send a human to the Moon.[5] NASA would not build its own towns, but would place its facilities in communities that already existed—that had their own governments, school systems, parks, road networks, houses of worship, and so on; these communities would grow together with the agency. One upshot of this was that, for the most part, NASA people were community people. That meant the culture and customs of the communities often became the culture and customs within NASA facilities.

As Julius Montgomery learned his very first day, and as countless others of the black men (and women) who came to take those white-collar jobs learned, the legacy of Yesterday weighed heavily on the Land of Tomorrow. There were old ways that everyone understood—subtle ways that dictated behavior. They stemmed from local history, questions of who lived where and why, who owned what, and who owned whom. That last part was particularly relevant in Brevard County; African Americans there were independent, entrepreneurial, and much less likely to stand by and take it, as Julius Montgomery had. They were that way, Theodis Ray said, for one reason—the history of the region. "There was a lot of history there," he said. "There was a lot of black people doing things that I don't see written anywhere."[6]

Laughing Waters in Allenhurst

For anyone who grew up during the Space Age the name "Cape Canaveral" holds an allure. Like "Paris," "the Pyramids," or "the Taj Mahal," it conjured an image. Cape Canaveral was the place where the countdown happened, the ignition and the liftoff. It is where hundreds of VIPs flocked to sit in bleachers in the sun and where thousands more camped on the beaches to watch the magical moment that was like nothing else in the world. Except for the launchpad, the most distinctive element of Cape Canaveral is the Vehicle Assembly Building (VAB) with its huge American flag and massive NASA logo painted on the side.

The VAB has not always been there, of course, and there is an inclination to think that all it supplanted was saw grass and swamps. In fact, where the VAB now stands there was once an African American town founded by freed slaves. It stood for generations, providing housing, jobs, and a home for black Floridians before the government wiped it out to make room for the construction of Kennedy Space Center. The name of the town was Allenhurst; Theodis Ray was born and grew up there.

The town's history began in 1872 when an African American family

named Campbell settled 180 acres on the Indian River Lagoon between Titusville and Oakhill.[7] The patriarch, Butler Campbell, named
his homestead "Laughing Waters."[8] Campbell was from Columbia,
South Carolina, and was one of about five hundred families that the
federal government resettled in Florida. The land was a reward, of
sorts, for helping in the fight for freedom. Officers of the United States
Colored Troops stationed at Hilton Head, South Carolina, in the fall
of 1865 organized the Florida Land and Lumber Company to start a
colony made up of freedmen and those friendly to them in Port Orange
near Mosquito Inlet, south of Daytona Beach.[9] Laughing Waters was
to be a place of black self-determination, where people newly freed
from bondage could toil and sweat for their own benefit for a change
and—for the first time ever—teach their children to read and write.

Campbell understood the power of education to keep his children
free, and others in Allenhurst felt the same, so, in 1890, Campbell and his
neighbor Andréw Jackson built what came to be known as the Clifton
schoolhouse for their children.[10] There is a historical plaque at the site
of the building indicating that another neighbor, Wade Holmes, "provided them with a one-acre lot on the northwest corner of his property" for the school. As was the case at many rural black schools at that
time, "classes were held during the summer months so the students
could help with the citrus groves." But even going to school only part
time, a black man known as Professor Mahaffey was able to teach the
children—these children of freed slaves—reading and writing along
with physiology, advanced hygiene, and, one year, Latin.[11]

Laughing Waters was even the site of one of the rarest of situations in the American South, where a white-owned company actually
bought land from black people. The Campbell family and the Jacksons
each had four acres of orange trees worth several thousand dollars. The
founder of Indian River Fruit eventually bought a piece of property
from Butler Campbell.[12]

The area languished in the first fifty years of the twentieth century, not unlike the rest of that part of Florida. The region was not a
pretty place. If people knew the area for anything, it was its "peculiar
climate so tropical for this far north in Florida and its plentiful mos-
quitoes."[13] That is how the renowned Harlem Renaissance novelist
and playwright Zora Neal Hurston described it in 1957. She lived out
her final years in a house trailer there among the orange pickers and

fishermen like Theodis Ray, who made up the region's economy. "As a young kid I worked where the VAB sits at today, packing fish and hauling fish into Titusville and Mims to be processed," Ray recalled. "Even before I had a driver's license, I was going out with my father." The area was not much, he said, but that does not mean there was nothing there. "We had life," he said. "Now don't get me wrong, the blacks didn't own land out there, but they had the orange groves—blacks did own a lot of orange groves. Plus they had businesses, too," like the ones started by the most prominent African American in the area, the man the black high school was named after, Andrew Gibson. According to Ray, Gibson "came out of slavery, too, but he set up what we might say today was the first little mini-mall. Because he set up the place that had a barbershop, a shoe shop, a café—all these kind of things. There's nothing written today about it."[14]

New Times/New Reality

That world ended on June 6, 1963, when NASA administrator James Webb wrote U.S. attorney general Robert Kennedy to say, "I have determined that it is necessary and advantageous to the interest of the United States to acquire certain land in Brevard County, Florida, for NASA Launching Sites for Manned Lunar Landing Program, Cape Canaveral, Florida."[15] The old fifteen-thousand-acre missile base that had been there for years was set to expand to eighty-eight thousand acres to accommodate Saturn V rockets to the Moon. There will be winners and losers in any boom and that was the case here, too. Hotel owners, real estate developers, builders, well-connected politicians—for them it was a bonanza. Not so for the fishermen and orange pickers, the people whose homes and means of existence lay in the way of those dreams.

Theodis Ray watched it all happen. "When the government came in and bought all this land, it wiped us out. I mean, you had these fishermen, that was their livelihood. Most of them didn't have education and all they did was fish."[16] The government demolished the buildings left behind, eradicating all evidence of the town with the exception of the Clifton schoolhouse (for reasons no one knows). In fact, "it survived along with some of its contents, such as an old trunk filled with letters, postcards, receipts and other Campbell family items."[17] As for the people, the government gave those who refused to leave meager compensation, but their livelihoods were gone.[18] The way of life that they had built there—African Americans working for themselves or for each other with the fruit of their labor feeding their community— was gone, too.

All that the former residents had left was the history, along with the pride and self-definition that went with it. Ray, like other former Allenhurst residents, took that over to the space center to look for a job. "After NASA come in and bought all that property," he said, "I started working out there as a janitor." He had driven a truck with a construction crew that was expanding the center, but once it opened, "the basic job that you could get back then was janitorial, or, if you could cook, you could cook. But as far as technical jobs [for blacks], no. They were just null and void." The situation was so desperate, he said, that "I worked with a guy that had a PhD, doing janitorial work. He just couldn't land a job down here, so he took what he could."[19] While that may sound apocryphal, it is worth noting that in 1964, Luther Hodges, the secretary of commerce in the Kennedy administration, said in a speech that "far too many trained Negro engineers are still sweeping floors, sorting mail, and digging ditches. Trained stenographers are working as maids. Trained electricians are raking leaves, and men trained as draftsmen can find jobs only as construction laborers."[20]

It is impossible to know whether NASA looked among its blue-collar staff for technicians, but it is highly unlikely.[21] It was looking elsewhere, however. Personnel from the Florida center had been in close contact with the placement officers of historically black colleges "to furnish the names of the engineer and science graduates, both new college graduates as well as alumni." During Theodis Ray's time there, "only three professionally qualified Negro applicants" had jobs at NASA.[22] Ray said NASA could have had more if they had wanted them, but he said the experience of people he knew who tried to get jobs at the Cape convinced him that NASA did not. "If you were an aircraft mechanic and had an A Tech license," he said, "if you came out to the Space Center and they had an opening for an A and a B, they would hire people of other races with a B into the A slot before they would put you there." If you were "a good ol' boy," he said, no one would ask about your qualifications. If you were black, "you need to know someone and be twice as smart as the average white." Julius Montgomery had a story that suggests African Americans were not even getting all of the menial jobs at Cape Canaveral. One day, he said, "I saw a white janitor! I stared. I could not believe it. I had never seen a white janitor in my life. I said, 'Oh God, this world is coming to an end.'" The word about workplace discrimination filtered through the community quickly enough, Ray said, and after a while, people stopped trying.

"No Formal Grievances or Appeals"

When he was younger, Ray hewed closely to his legacy as the descendant of people who had fought for their freedom. "When we have an issue that we can't resolve one-on-one, we protest," he said.[23] He did that in the community, joining a national trend of sitting-in, and helped close down a local ice cream parlor and gas station that refused to serve blacks. The loss of his community seemed to break his desire to fight against discrimination at NASA, however. After a few months sweeping floors at Cape Canaveral, he had had enough. In 1964, he took his pride and his initiative and went where he thought people would better appreciate him, enlisting in the U.S. Marine Corps. In the period before he left, Ray did not sue or try to file a grievance of any sort. There is reason to believe he is not alone in that. According to a report by the personnel office for Cape Canaveral in 1963, "since the establishment of the [space] Center, no formal [civil rights–related] grievances or appeals have been filed." It went on to elaborate that "no complaints of discrimination have been filed with the deputy employment policy officer and there are no indications of discriminatory or segregative practices."[24] Certainly, Theodis Ray was not the only African American facing discrimination or feeling discriminated against, so how could that be?

Remember that NASA people were community people. It is quite likely that the white NASA staffer who wrote the memo was unaware of the reality of the situation.[25] An agency-wide survey of NASA's African American employees that same year found that while 77.8 percent "felt that EO [equal employment] counselors are fair and objective," and 86.8 percent of blacks "had considered using" the agency's Equal Employment Counseling program, nearly half (46.4 percent) said they lacked "faith in the counseling or complaints system."[26] African American workers at NASA "almost without exception, were suspicious of the EEO policy." Black NASA people, who were community people, too, "were not convinced that the policy would be carried out." They knew their history and viewed their world from its perspective. So when they would apply for jobs at NASA, they would "notice that Negroes are not employed in the employment office. They are convinced that they have less chance of passing civil service examinations because white people read their papers."[27] This was likely also the case when it came to filing grievances. After his tour in Vietnam, Ray said NASA "offered me the [janitor's] job back." But now he was an Allenhurst resident and a Marine. "I was too evil to take that job," he said. "Most of the people that govern you in janitorial was uneducated and you had more experience than they did, but they was in charge so you had to go along to get along if you want to keep your job."

Knowing Your Place

During the early 1960s, NASA searched desperately to find African Americans to move south to work at its facilities. "Statistics indicate that NASA is at the bottom of the list among government agencies in the employment of Negroes," Administrator Webb confided in letters to his center directors in 1961.[28] He lamented two years later that "the Vice President has expressed considerable concern over the lack of equal employment opportunity for Negroes" at NASA.[29] A report the following year said the agency was "still behind in government-wide ratios of Negro to total employment in both the classified and wage board areas."[30] Desperate to find people, NASA sent recruiters out around the country asking African Americans to come. Julius Montgomery was one of them for a short time. "I could not get any black people to come," Montgomery remembered. "I called up, and I talked to a fella and he said, 'Do what? Come South?! Are you crazy?'" People were afraid to come, he said, "because of all the things that were going on—the bombings, the killings, the shootings of the people on the buses. All of that played a part."[31] There was another problem, and it was one that Theodis Ray said was vital.

Anyone trying to grasp this problem, he said, needs to understand the nature of racism as it existed in the middle of the twentieth century. In the South, African Americans were often reluctant to cross state lines because their very survival was so dependent on knowing the local customs—knowing who was going to protect you and who was not. It speaks to a reality pointed out in a lesson African American parents have passed down to their sons for years in a ritual known as The Talk. "Okay, you're black," Ray remembered his father telling him, "and you are going to be ostracized because you are black." Your life has no value to the white man, his father told him, and if you remember that, it will keep you alive. They will call you a "coon," or worse. They will treat you as if you are nothing. "Look past this anger when they are going to call you these bad names," his father told him. Even though they were from Allenhurst and "nobody controlled us, we just understood what we were and our identity and to not try to venture beyond that," he said. Pushing back against local Allenhurst history was the legacy of racism in the broader nation. Because of that, he said, "We knew when to say, what to say, how to say, and who to say it to."[32] That state of being—the fear and diminishment with overtones of practical subhumanity—was not just the lot of the black Florida fishermen or black pickers in the orange groves, of course. It was the fate of the most vaunted members of the community. It was the fate of Harry T. Moore. Terror compounded over centuries can put an entire race of people, even the proudest members of that race, into a defensive crouch.

That is behind Ray's explanation for why NASA could not get African Americans to move south or to get blacks in the South to move to where the jobs were. "Being black, each state has a preference. Me being raised in Florida, I wanted to stay in Florida. I didn't feel comfortable being in Alabama." A black Floridian in Alabama, he said, had no idea which white man held the power to decide whether he would live or die. In Alabama, Ray said, "Governor Wallace may tell you you're one of his boys; he's going to say to you, 'Boy, if you can keep yourself out of the graveyard, I'll keep you out of jail.' So when you're in Alabama, if anything should happen—a state trooper run down on you; whatever it might be—you call Governor Wallace's boy and you're covered. Now you come to Florida and it's different. You come to Georgia and it's different, because you're not under that blanket anymore." This was a fact of life for blacks throughout the South, Ray said. "It doesn't sound logical, but it's the truth. Each state, they had a certain way they treated black people."[33] William Bell, who worked at Cape Canaveral in the early 1960s, said the same thing. He said the reality came home to him one night after a dispute between two black men where someone called the white sheriff. Both parties had legitimate claims, but Bell said the sheriff resolved things by saying, "Well, John here is my nigger, and if John says that's the way it happened, that's how it happened." John went free. The sheriff hauled the other man off to jail.[34] This is what recruiters were up against as they tried to lure African Americans to join the NASA job rush to eastern Florida.

The Space Race and Race

The action that necessitated all these changes—that caused Cape Canaveral to expand—was, of course, the decision by President Kennedy to land an American on the Moon. The president announced that decision in May of 1961. Many events made news that May. Harper Lee won the Pulitzer Prize for *To Kill a Mockingbird* and Fidel Castro— just days after crushing the Bay of Pigs invasion—declared an end to elections in Cuba. Those both happened May 1. But as things would turn out, it was not to be a month preoccupied with the Cold War or the arts. May of 1961 goes down in American history as a watershed in two areas: civil rights and the American space program.

This was the month when Alan Shepard became the first American in space. It was also the month when President Kennedy, playing catch-up against "the Soviets with their large rocket engines," who just days earlier had put their first man in space, made his audacious promise that by the end of the decade, America would land a man on Moon and return him safely to the Earth.[35] May 1961 was also the start of the Freedom Rides, where young people placed their lives and personal safety at risk to challenge the horror and violence that helped keep segregation in place.[36] Nine days after Shepard's historic flight, seven black and six white young people got on a bus in Washington, DC, headed for New Orleans—whites in the back seats, blacks in the front.[37] Ten days later, their bus was firebombed in Anniston, Alabama. A week after that, as the Riders moved on, mobs rampaged and rioted in Montgomery, forcing the governor to declare martial law.[38] This mayhem—and more that followed—drew the Kennedy administration, "like it or not, in a struggle with Alabama and other southern states over the power of the federal government to protect the civil rights of African Americans."[39]

By the time June rolled around, the United States was committed to going to the Moon and the Kennedy administration had jumped with both feet into the fight over civil rights. There is sufficient evidence that Kennedy did not take on either of these two missions voluntarily, despite efforts at the time and in the years since to suggest otherwise. Numerous civil rights histories say Kennedy once defensively claimed that he had "done more for civil rights than any president in American history."[40] Historians take a dimmer view when it comes to assessing not only Kennedy's interest in the civil rights cause but also his administration's success or failure in accomplishing anything in the field.[41] In his speech following the confrontation over the integration of the University of Alabama, according to one historian, "no president since Lincoln had addressed the nation in such forthright moral terms about American race relations."[42] It is also true, however, that when the Supreme Court ruled to allow James Meredith to enroll at the University of Mississippi, Kennedy told the state's governor Ross Barnett, "This is not my order. I just have to carry it out."[43] It is also said that in the area of policy, Kennedy's actions "lacked passion and often seemed cynical and apologetic,"[44] and that overall he "proved to be more style than substance on civil rights."[45]

As a politician, Kennedy was a pragmatist, so his party affiliation made the civil rights problem particularly vexing. The South was, after all, home to the most entrenched and powerful element of his base. Since the 1930s, southern politicians in Washington had obtained and held on to power using a congressional seniority system that rewarded longevity. They had longevity because they came from states where significant proportions of the population were not allowed to vote.[46] It was a vicious circle: laws and customs kept blacks from southern voting booths; whites elected the same senators and congressmen over and over; they rose to positions of power through seniority, which they used to continually choke off legislation that could have given African Americans in their states the opportunity to improve their lives through access to the ballot box. Any president who would stand up to these powerful men was risking his own political survival.

As with civil rights, there is ample evidence that going to the Moon was not a priority for President Kennedy, either. He certainly did not talk about an end-of-the-decade Moon landing during the 1960 campaign.[47] In fact, Kennedy made his outer space speech as an act of desperation. His presidency was failing in May 1961 and he had to do something.[48]

Part of the Cold War was a constant argument over which economic system—capitalism or communism—emerging states should adopt. There were twenty-five new African countries in May 1961 and eight more were on the cusp of organizing.[49] The decision about which superpower they would align themselves with was up for grabs.[50] With the launch of *Sputnik* in 1957, the crash landing of the Soviet craft *Luna 2* into the Moon in 1959, and the launch of Yuri Gagarin into space only six weeks before Kennedy's speech, the U.S. was badly behind. Kennedy had come in promising to get the country moving again, but by May of 1961, the Soviet Union had become the first nation to put a human into space. Shortly after, when America had its disastrous experience at the Bay of Pigs, the Kennedy administration was back on its heels and the Man on the Moon speech was a Hail Mary pass, an attempt to change the dynamic.[51] As presidential historian Michael Beschloss wrote, "On April 20, the day Kennedy knew for certain that the Bay of Pigs had failed, [Kennedy] called in Vice President Johnson, chairman of the Space Council, and asked him to come up with something fast in space."[52]

So it was in May 1961 that the administration took on the dual missions of equal rights and space exploration. Whether it threw itself in or was thrown in, once in, it pursued both with equal amounts of (to use the president's favorite word) "vig-ah."

Pride, "Tinged with Growing Bitterness"

Putting a man on the Moon meant spending lots of money. The first firm estimate of what it might cost to go to the Moon came in 1964, when it was put at $19.5 billion. For perspective, aid to elementary and secondary education that year was $2 billion. President Johnson's War on Poverty annually was $1.8 billion.[53] Much of the money spent on NASA went for hardware, but most of it went to hire people, both employees and contractors. There were astronauts and the people at Mission Control, of course, but also people to construct NASA facilities, calculate trajectories, solder wires, drive rivets and install phones and computers, serve meals, conduct medical experiments, write press releases, cut the grass, and so forth. One result was a population explosion in Florida.

Brevard County's 1960 population was 111,435 people, 98,909 (88.7 percent) whites and 12,526 (11.2 percent) blacks. The city of Cocoa, with a population of 12,294, had 3,823 (31.1 percent) nonwhite residents, the largest nonwhite population in the county. Cocoa was part of the Cocoa-Rockledge division of Brevard County, which had a total population of 15,775. By 1970 the county was home to 230,101 residents; 208,389 (90.5 percent) whites, 20,689 (9 percent) blacks, and 948 (.4 percent) other. The Cocoa-Rockledge division grew to a population of 32,982, of whom 5,360 (16.3 percent) were black.[54] From 1960 to 1970, the black population of Brevard County increased in absolute numbers by 8,163, a 65 percent jump. This increase drowned in the flood of 109,480 whites entering the county, a +52.5 percent change from 1960.

To accommodate the population boom, the Brevard County Housing Authority took over what had been predominantly black neighborhoods in the town of Cocoa to build public housing—housing that Theodis Ray pointed out was "all white-only." County authorities urged African Americans to resettle in an outlying district, "where, incidentally, they would be ineligible to vote in municipal elections."[55] That area was also a considerable distance from Cape Canaveral and its jobs. A reporter named William S. Ellis visited the space communities in the early 1960s to write a series of articles called "Space Crescent" for the *Nation*. In his article looking at race issues in the space communities, he called the region where the black community had been moved a "snake swamp."[56] Many of the evicted families earned too much money to qualify for public housing (they had been living in "comfortable homes" before the government took their land, Ellis said) and faced two choices: take lower paying jobs to qualify for public housing or live in their cars.

This entire situation came about so that the community could accommodate NASA, and with all the complaining that the agency did about not being able to lure African Americans, it would be reasonable to expect NASA to take a stand against it. After all, if NASA was "the most logical employer to initiate the kind of programs necessary to achieve true equal employment opportunity," how could it not? So, did NASA step in to stop housing discrimination? Sadly, it did not. "We have not entered into the social aspects of integration here," an agency spokesman told Ellis. NASA's goal was "to put a man on the moon," he said. "That is what we are concerned with."[57] The Reverend W. Oliver Wells, who had taken over for Harry T. Moore as head of the area's NAACP chapter, confirmed this. Reverend Wells was the one who mentioned the people sleeping in cars. He also said that while there were African American police officers in the community, they could not make arrests—they had to call in so the station house could send over a white officer to do that. "What progress there is has come through the efforts of local people," Wells told Ellis, not from NASA or the Air Force.[58] Ironically, Reverend Wells said, discrimination was another reason NASA was having such a tough time getting blacks to move down there for work. "Even if there are skilled Negroes available," he said at the time, "they will not come here because there is no place for them to live." He said he knew an African American General Electric worker in the northern part of the state who wanted to transfer to Cape Canaveral. "When he learned of housing conditions for Negroes in this area," he said, "he decided not to come."[59]

Roy Wilkins, the executive secretary of the NAACP, joined the protest against government inaction in Cocoa. In letters to the President's Committee on Equal Employment Opportunity (PCEEO); to David Lawrence, chairman of the President's Committee on Equal Opportunity in Housing (PCEOH); and later in a letter to the president himself, Wilkins submitted affidavits complaining of racial discrimination in Cocoa in both employment and housing insured by the Federal Housing Authority (FHA) and the Veterans Administration (VA). He also submitted charges of workplace discrimination at Pan American Airways, a major government contractor at Cape Canaveral. The affidavits, which Wilkins obtained from the NAACP's labor secretary, Herbert Hill, charged that Pan Am had denied two African American employees opportunities for "upgrading [professionally] solely because of their race." A third person complained that he could not get a job at Pan Am in the first place because he was African American. Wilkins asked that the government end workplace discrimination and discrimination in public housing projects. The Cocoa–Cape Canaveral area was of particular interest, he said, as a community "where on an existing pattern of segregation, extensive residential expansion is being planned with federal assistance." The *Baltimore Afro-American*, the *Atlanta Daily World*, and other black papers gave Wilkins' actions front-page coverage.[60]

Wilkins' letter to President Kennedy was sharply critical. He charged that extensive federal involvement in the Cocoa–Cape Canaveral communities warranted special attention to discrimination. He cited the local economy's dependence upon space and military installations. He also talked about how the government used federal money to help the communities deal with rapid expansion. In addition, the letter took the popular tack among African American opinion leaders of the time, weaving in a telling allusion to the myth of the Space Age and its failure to deliver the promised better world. Wilkins wrote:

> There is a particular irony when the soaring aspirations exemplified in the United States government's programs for probing the far reaches of space are contrasted with the harsh reality faced by Negroes who are contributing to those programs but find themselves barred from decent homes whose construction was made possible by government aid. There is a sharp contrast between the world leadership we seek in space achievement and the handicaps resulting from job discrimination which keep us from the full use of the manpower needed for the effort.
>
> There are approximately one thousand Negroes employed at Cape Canaveral and at Patrick Air Force Base nearby. These employees speak with pride of their participation in these vital activities, but the pride is tinged with growing bitterness at their government's failure to ensure equality of access to homes, job advancement and schooling for their children. It is not only their deprivation but the nation's shame, for example, that as many as 300 FHA-financed homes stand vacant and unused in the Cocoa metropolitan area, while Negro workers at Canaveral pass them en route to work from their substandard homes, furnished rooms and trailers.[61]

Wilkins did not call directly upon NASA to resolve the housing problems in Cocoa; in fact, he made no request of NASA at all. In preparing a response to Wilkins, Lee C. White, assistant special counsel to the president, sent a memorandum to NASA administrator James Webb, along with memos to the Department of Defense, the Housing and Home Finance Agency, and Hobart Taylor, executive vice chairman of the PCEEO. PCEOH chair Lawrence mirrored Wilkins' complaint, saying, "Surely there is something tragically paradoxical when the men and women who are making space exploration possible are denied access to housing in the very shadow of the launching pads at Canaveral." He promised to use his agency's "good offices to correct the situation." Taylor, meanwhile, promised that the PCEEO would give the employment problem at Cape Canaveral "continuing study with the view of initiating appropriate affirmative action." Webb for his part answered that NASA had no record of any complaints about discrimination in the Cocoa-Canaveral area. Furthermore, because neither housing nor public facilities used NASA funds, he said the agency's powers of direct intervention were limited.[62]

Theodis Ray was one of the people Wilkins referred to, whose "pride [was] tinged with growing bitterness." Back from Vietnam in late 1965, he found himself swept up in the Cocoa housing crisis, which had not abated despite the work of David Lawrence's "good offices." Ray immediately found a job doing logistics work at the Cape for the NASA contractor LTV. He was soon a "lead storekeeper," the manager of a warehouse where NASA crews could get "anything that you need from a pencil to a missile component."[63] Despite the fact this was a good-paying job, his family was one of the ones caught in the housing squeeze. "We just couldn't have decent housing," he said. This was another example of "what they called back then when they want to scatter you about, 'Urban Renewal.' Every time they mentioned the words 'Urban Renewal,' there we go again. You've got to give up your school. You've got to do this. You've gotta bus here, you gotta bus there. It never worked for us."[64]

While many protested the housing discrimination, Theodis Ray did not. It was not because he feared losing his good job. "That never bothered me," he said. Instead, he stayed out of it because, during his short time away in Vietnam, the nature of protest had changed. In the early days, he said, "the NAACP taught you how to conduct yourself." Adherents to Martin Luther King Jr.'s principles of Gandhian nonviolence, the older protest organizers believed in dressing up, being quiet, applying soft pressure, being tenacious, and winning through attrition. That ethos went by the wayside after the Watts riots in August. Protests after that "got more serious," he said. "I stopped marching because they got too violent." First, the marchers turned up the volume, he said, and then the police followed suit. "They turned dogs loose on you. They used water hoses and I can't deal with all that kind of crap." So he stopped. "Some of them got really bad—really violent. I got out of those."[65] Instead, he quietly seethed, complaining to anyone who would ask. The housing situation gnawed at him ("I lived twenty miles away from NASA," he said, and "you had to drive every day"), and he was clearly bitter about all of the discrimination related to his and his community's experience with the space program. His living situation was not the one he had expected. He was not the only one who felt that way.

The National Academy of Arts and Sciences had conducted a study earlier in the decade premised on the idea that NASA might present "a new era of equality according to ability."[66] Places like Cape Canaveral, "where the lives of the people are dominated by aerospace activities," they reasoned, "might in some way represent a microcosm of the future." Instead, they were disappointed to find that the idea that the Space Age would "open up opportunities for all" was merely a "myth of 'the future.'"[67] Theodis Ray reached the same conclusion, and that was what left him most bitter. He had hoped that this new world of tomorrow with human beings blasting off into space would "change the way we live and make it a better place for us to grow up." Instead, "integration never panned out the way they was signed up to be. Integration to me, I thought was going to be for us to get better jobs," he said. Instead, all it did was allow white people "to be a little smarter and use us in a different manner."

Dissatisfied with the present, Ray did what he had always done; he reconnected with his past. "Our parents always taught us how to handle yourself when this or that happens," he said, and one of the principal lessons they taught was to lean on the past when the present was grim. "We were proud. We knew what we were." Perhaps as a balm to alleviate the sting of injustice, he found himself reminding people about the source of that pride. While the people at the place he called the "historical joint over there on the corner on the main street in Titusville [the North Brevard Historical Museum]" did not recognize the importance of Allenhurst, other people were beginning to. The North Brevard Heritage Foundation began asking around about Harry T. Moore. That brought Ray into contact with Roz Foster, an amateur historian at the foundation who started pulling groups together for oral history sessions that delved into the past to flesh out the full story of Allenhurst.

The sessions acted like a sort of therapy and slowly, through the bitterness, Ray's pride began to shine through. "Just to watch Alan Shepard go off into outer space," he said, "it felt like we were doing something worthwhile with our jobs out there. And then, actually, when we landed on the moon and '69, man, it just seemed like it was a new heaven opened up!" Despite everything, the space program, he said, was "just great. It's just great to be a part of it." Above all, said the man from Allenhurst—the man who knew where he came from and knew why it mattered, "in particular, it's great to be in Florida, and I am proud to be who I am."

The idea that the federal government was "the most logical employer to initiate the kind of programs necessary to achieve true equal employment opportunity" was popular at the highest echelons of the Kennedy administration. As they looked for ways to ramp up that pride and tamp down the bitterness, they struck on a plan. Their idea was to pump new money into portions of the country where segregation was at its worst, to use that money to create new kinds of jobs that African Americans could hold, and then to create new rules to force African Americans into those jobs. It was a solid plan and they put an enormous amount of effort behind it. The experiences of those who lived and worked through it reveal all that the planners faced. They could plan and they could spend, but on both sides of the integration fight people were complicating the challenge, and only people would be able to work through it.

Stronger Than Steel
Frank Crossley

My fear gene seems to be very weak.

FRANK CROSSLEY

As an engineer, Frank Crossley created metals that were stronger than steel but much lighter. The irony is, doing that as an African American in the 1960s required a fortitude that was stronger than steel, but with a touch that was much lighter. Brilliant, but with an unfailingly self-effacing personality, Crossley started out fast and rose to positions of authority in the military, the world of science, and the corporate suite that were so unprecedented that he was regularly vaunted in the pages of the black press. He battled prejudice throughout his life: denied food because "we don't serve Negroes," denied entrance "because they did not admit Negroes." He ended up studying metallurgical engineering rather than his first choice, aeronautical engineering, because the school that taught aeronautical was too far to travel to without going through neighborhoods where police would tail him and the residents accost him with kicks, spitting, or worse.[1] He recalls supervisors at two different companies telling him he had not received a promotion because he was "so advanced for a Negro [that] we thought you were content."[2] It would be wrong to say that it did not bother him when he saw his progress stunted by the wider world's prejudice. It clearly did. Even at the end of his life, he spoke about each slight with disdain. However, while these experiences bothered him, he never let them get in his way. He seemed always to face them by rolling his eyes, slowly shaking his head, and then continuing to excel.

So many years removed from the battle over workplace integration in America, it is natural for the perception of that time to be somewhat unclear—for the lines between "the unrecoverable past and our memory of it" to blur.[3] In cases like this, a popular inclination is to default to the "accepted paradigm" that shapes our thinking from the distance of years.[4] Within that paradigm, the tendency is to look at workplace equality and ask why it took so long.[5] The story of Frank Crossley helps explain instead why it happened as quickly as it did. His experience fuses a personal and a national narrative at the core of understanding the integration that took hold in American factories and offices following World War II. While Crossley was never a NASA employee, the work he did developing metals that became the skins of rockets and missiles was vital to its success. His experience is also indicative of the changes the space program helped bring to race relations in America during the early 1960s.

"One Clean-Cut and Fine-Looking Negro"

Crossley grew up in the 1920s and 1930s on the South Side of Chicago. His father was a railroad porter and his mother worked in a button factory. "The United States in the 30s, for a black person, could be a very hostile place," Crossley said.[6] In particular, job prospects for African Americans, as we have made clear, were severely limited, and his family looked to give him every edge he could get. Crossley had three uncles with college degrees. The one who was a lawyer encouraged Frank to be a doctor, telling him, "You don't have to worry about being employed by anyone." Medicine was not what young Frank was interested in, though. "Engineering appealed to me," he said.[7]

He completed two years of study at Illinois Institute of Technology (IIT) when World War II interrupted his studies. Crossley's father had been a sergeant in the infantry in the First World War and Frank had been an ROTC commanding officer in high school, so, with the war on, he entered the U.S. Naval Reserve Midshipmen's School, which trained naval officers. He was the only African American in a class of 1,500, a fact that got him attention—the type of attention that stayed with him much of his career. At the time, S. I. Hayakawa, a future U.S. senator, was writing a column for the *Chicago Defender*. He took note of Frank

Crossley and wrote several pieces about him. "Among all those clean-cut and fine looking white boys," a typical column ran, "there is one clean-cut and fine-looking Negro student. Residents of the neighborhood have probably wondered who he is, what he is doing there, and how he is getting along."[8] Their wonder, and any concern they might have had, was justified.

The United States Navy has a complicated racial history. While it was the case for years that there was not a service of the U.S. military more segregated than the navy, the navy was in fact the U.S. military's first proving ground for integration. Fifteen hundred African American men served in the navy during the Revolutionary War, and a black fife player and seaman named Cyrus Tiffany helped save the life of Commodore Perry during the War of 1812. From the 1830s through the 1850s, about 5 percent of the U.S. Navy consisted of men of African descent. At the same time, the number in the U.S. Army was zero.[9] By 1864, more than 25 percent of the U.S. Navy was black. After the Civil War, however, as Jim Crow took hold, harsh and vivid discrimination gripped the navy.[10] Whether that discrimination was worse than in the army is difficult to say, principally because of the nature of military service. Armies exist on land and can peacefully segregate, as the U.S. Army did. There is not enough room to do that on a ship, however, so the navy segregated by job category. For more than seventy-five years after the Civil War, African Americans on navy ships were limited to being cooks, assistants to cooks, and valets for white senior officers.[11]

First Lady Eleanor Roosevelt, among others, pushed for conditions to change in the U.S. military. In 1940, she suggested that one way to move toward integration was to have all-black marching bands in the navy.[12] The B-1 Band in North Carolina and two other bands in California and Michigan allowed African Americans to hold navy ranks other than cabin boy, and by 1944 Mrs. Roosevelt and Adlai Stevenson had persuaded the president and the navy secretary to go even further. As an experiment, the navy commissioned a handful of African Americans to see if white crews would take orders from black officers. The black press dubbed the first group of African American commissioned and warrant officers in the United States Navy "The Golden 13."

As far as Frank Crossley was concerned, the impact of even this token integration was remarkable. He and the white cadets, he told Hayakawa, "do everything together. We work, study, drill, eat, march, and live together." He did mention the bad part, but he played it down. "The only time I don't join the rest of the boys is when they go out on social engagements," he said, but "that doesn't worry me because I have friends of my own in the city; besides, I'm working too hard to go out much."[13] Crossley received his commission as an ensign in November 1945, two months after the war ended and about eighteen months after the "Golden 13." Only one of the Golden men ever got a ship assigned to him. The navy thought so much of Crossley, however, that it put him in charge of the second deck division—the sailors who do the fueling and painting—on a troop transport ship. Crossley had thirty-six men under him, including an officer. "When they first saw me, they looked surprised—the way the boys at Illinois Tech look at you when they find that you are going to be their English teacher," he said at the time.[14] But as he would throughout the rest of his career, he handled the stress of command and the potentially dangerous novelty of the racial situation onboard his ship with calm and expertise. "I think there are two generic characteristics that I have based on my observation of other companions who were my age," Crossley said. First, "My fear gene seems to be very weak." Second was something he learned when he was very young. "An older friend—and I was still in elementary school when this happened—basically taught me to control my emotions," he said; particularly anger. "I used to be hot tempered," Crossley said, "and he taught me control." This led Crossley to take a radical approach when he stepped onboard his first ship. "When I joined the navy," he said, "I decided that if I heard the N-word and it was not directed to me—I simply overheard it because it was commonly used in those days—I would not take offense. I would only take offense if it was directed to me."[15]

When Hayakawa interviewed Crossley at the end of the war, Crossley said he agreed the experiment was a success.

"Did your men show any resentment ever being commanded by a Negro officer?" I asked.

"No," said Crossley.

"Not even a little bit?" I asked again.

"No."

"Well, did they ever notice that you were Negro?"

"Yes," Crossley said. . . . "[But] after they got over their surprise, they acted towards me just as they do to other officers."[16]

In Crossley's wartime experience, the fact that he returned home with "no dramatic stories of injustice, of fights against prejudice, of indignities nobly borne, of triumphs over discrimination," Hayakawa saw a tantalizing harbinger of the postwar world. "His uneventful and untriumphant tale is itself a triumph," he said. "His kind of uneventful, peaceful integration in joint work with other men is the kind of thing that we hope for all over America." This Pollyanna vision was expressed in the euphoria attendant at the birth of the "American Century"; but Hayakawa did hit on something. There turns out to be an element in Crossley's story that would in fact be present throughout his career. "The important thing about the lack of integration," Hayakawa said, "is that it doesn't give people the chance to be human to each other, and encourages instead the spread of superstitions about the 'other' race."[17] Crossley always strived to stay as close to white people as the law and custom would allow. He did this with a purpose, always mindful that he could prove by example the absurdity of superstitions about the "other" race.

Johnson's Plan to Change the South

After the war, Crossley was ready to start a career. He took a teaching job at Tennessee Agricultural and Industrial State College (A&I), the all-black college now known as Tennessee State. His timing could not have been better, as this was the period during which Lyndon Johnson was driving the idea that government could bring about economic and social progress by spending enormous sums of money.[18] In the 1950s, Johnson and other southern legislators in Washington were looking to force economic progress in the states of the Old Confederacy. Whatever the end game was for the other southerners, for Johnson racial harmony was part of the package.

Like many other southern New Dealers, Johnson thought that the roots of injustice in the region had grown thick in the soil of southern poverty.[19] Hatred and discrimination persisted because people were all fighting over the same meager economic pie. Fix the economy, Johnson believed, and segregation would crumble away. This was a job, he felt, which could not fall to the region's private sector. He saw southern businessmen as too invested in keeping their workers—both black and white—languishing in ill health, hunger, and ignorance. The South at that time was, as economist Gavin Wright wrote, "a low-wage region in a high-wage country."[20] As such, its business leaders and the local politicians they controlled lived in constant fear of anything that might cause folks to up and move away. This was not an idle fear; they had seen it numerous times. It happened first when emigrant agents enticed blacks away from Alabama, Georgia, and the Carolinas to higher paying jobs in Mississippi, Arkansas, Louisiana, and Texas between the Civil War and World War I. It happened again during the Great Migration, when the proportion of African Americans leaving the South entirely grew from 20 percent to more than 50 percent. They could not see it continue and felt they had to keep outside influence at bay, which is why outside influence was precisely what Johnson sought to deliver. Bringing in the full force of federal power was the only way he saw of integrating the South into the mainstream of America's economic life.[21]

When W. J. Cash talked of there being "many Souths," he meant that the Old South had its timber, its coal mines, and its mills.[22] The monumental changes that precipitated the 1960s-era intervention known as the "Second Reconstruction," however, came about principally due to the death of one particular "South" that had dominated the region's legend as much as it did its economy.[23] The system of cotton-based sharecropping had been under pressure since the beginning of World War I, and, by the time of the Great Depression, it was on its knees. In the end, cotton acreage dropped from 43 million in 1929 to less than 15 million by 1959.[24] By the 1930s, the situation had become so bleak that the region's power structure felt it had no choice but to accept a dose of federal help, regardless of the consequences.[25] Taking federal money meant accepting both policies that flattened wage differentials and workplace protections that kept northern jobs from "heading South."[26] It meant accepting cutbacks on cotton acreage. It even meant limited enforcement of policies banning race discrimination.[27] As time went on and World War II showered the region with military bases and federal contracts, the South of old faded further.[28]

By the 1950s, the war contracts were ending. The military bases were scaling back or closing. What might take their place and continue the federal intervention to the point where it might actually bring about racial harmony? Lyndon Johnson had an answer: spending on space.[29] As he saw it, space was a federal jobs program (something his more recalcitrant southern colleagues might balk at) that he could tie to American prestige and to fighting the Soviets in the Cold War.[30] The jobs it would create would be high-paying ones—jobs that would lift all southerners, black and white.

Stories of African American NASA workers and those who, like Frank Crossley, made their way elsewhere in the space program suggest Johnson was on the right track. About a year after *Sputnik* launched, Congress passed the National Defense Education Act, and more and more American young people began to think about careers in engineering and science. According to James Jennings, a retired NASA deputy administrator, however, that movement bypassed the black community completely. "We heard *Sputnik* on the radio," Jennings said, "but at that time we didn't even dream of being able to participate in it. I mean, most folks went to school to be something in the black community. They didn't think about that they could go out and actually be a part of something like a space program."[31] Other African Americans who ended up at NASA shared that bleak assessment. While Frank Crossley felt otherwise, he could not help being realistic. "I decided that I would study engineering," he said, "and if I couldn't get a job—and I thought my chances of getting a job as an engineer in the United States was less than 50 percent—I decided I would either go to Canada or Mexico. Canada had the virtue of speaking English and Mexico had the virtue of having colored people."[32]

As Johnson envisioned, when the space program came along, the job dynamic changed. Even a low-level space job like a "computer" was a desk job, where a person worked with his or her mind instead of his or her back.[33] There is no question that NASA's black employees shared a sense of the agency's mission. What was more important, however, was that for African Americans struggling to achieve economic as well as social equality in the face of Jim Crow, NASA offered a way up and a way out. Along with a living wage, NASA also offered job security. All of that was a change from the lives southern blacks lived before Kennedy announced America was going to the Moon.

As Kennedy was making his Moon decision, Frank Crossley was making decisions, too. He had tired of Tennessee. The president of A&I, he thought, "came through to me as just a flunky for the white power structure." Therefore, he traveled back to Chicago and back to IIT, where he made contact with Kurt Peter Anderko, one of the German scientists the United States brought here after the war to keep them out of the hands of the Russians. Anderko was the head of the metals research department at the IIT research institute. Crossley settled in at IIT, working on metals for the emerging aerospace industry as the Kennedy administration came into office.

Equal Employment Opportunity

While NASA pursued the initiatives of the Kennedy and Johnson administrations to promote equality, it would be wrong to give the impression that civil rights began with Kennedy, though there are certainly those who would like to suggest it. The movements for desegregation and voting rights rushed ahead with considerable force during the Kennedy years, but even before then, many, including Julius Montgomery, benefited from initiatives of the Roosevelt and Truman administrations. During the Eisenhower administration, presidential executive orders, the Civil Rights Act of 1957, and action by government, business, and private social and religious groups, both black and white, made additional progress.[34] Roy Wilkins and Martin Luther King Jr. praised Vice President Richard Nixon, a "card-carrying member of the NAACP," for his work to pass the 1957 Civil Rights Act.[35]

As for why Eisenhower got as engaged as he did with the civil rights cause, a consensus opinion is that the credit goes to international reaction to demands and protests by African American communities and their representatives. A thoroughly unintended consequence of the decision to base the headquarters of the United Nations in New York was that the move brought African Americans a worldwide stage for their plight and their demands to address it.[36] America's enemies abroad, who were anxious to show newly emerging nations in the Third World that Communism, not the American free enterprise system, was the better way to achieve worldwide harmony and understanding, egged them on. Discrimination made America's allies uncomfortable, too, and as large numbers of UN members, including America's friends, began attacking the U.S. for its racial policies, Eisenhower realized that change had to come.[37]

As the Kennedy administration came into office in 1961, civil rights protests were already beginning to simmer and the president was pushed by the liberal wing of his party (which had inserted a strong civil rights plank into the platform at the party's 1960 convention) to take action on equal rights. In addition, the pressure Eisenhower had been receiving from overseas continued.[38] Kennedy knew—and he told allies black and white—that southern committee chairmen would guarantee any civil rights bill would be dead on arrival in Congress.[39] He also recognized the political danger that came from threatening to withhold federal funding to states that discriminated; holding back money for highway construction or aid to hospitals not only risked angering Democratic southern governors but also the aid recipients themselves and their lobbyists and activists.[40] Still, the pressure was on and he felt compelled to respond, so the president decided to fight the battle for racial equality on two fronts he knew he could control. The Justice Department would tackle the desegregation of public accommodations and schools, while equal employment opportunity was addressed through the hiring provisions of federal contracts overseen by the Labor Department and enforced by the President's Committee on Equal Employment Opportunity (PCEEO).[41]

Kennedy's labor secretary, Willard Wirtz, later said, "The larger edifice of civil rights is itself dependent on the quality of employment opportunity. Without it, the quality of citizenship is only an empty phrase."[42] That is not why the Kennedy administration concentrated on federal workplace integration to the initial exclusion of other equal rights, however. They did it because it allowed the president to address the problem on his own—to demonstrate action on civil rights without having to work through Congress. He preferred the approach espoused by Martin Luther King Jr., who said, "The President could give segregation its death blow through a stroke of the pen."[43] The political realities of the Kennedy years dictated that the president turn to his executive powers, rather than legislation, as his principal means of addressing discrimination, and on March 6, 1961, Kennedy did just that by signing Executive Order 10925.

The order was a three-part plan to review federal employment practices and promote equal employment opportunity within and by the executive branch of the federal government. Part 1 created the PCEEO. In part 2, the president authorized the PCEEO to recommend changes in federal hiring practices and directed all executive agencies to examine their employment policies. The order, in part 3, required all government agencies that contracted work out to the private sector to include antidiscrimination clauses in their contracts and apply antidiscrimination policy to their contractors and subcontractors. It also told the agencies to provide nondiscrimination recommendations for those service providers' labor unions. Federal contractors had to identify themselves as equal opportunity employers and declare in their advertising that they would not discriminate based on race. Companies that did discriminate risked losing their contracts.

Kennedy tasked Johnson with drafting EO 10925, so it makes sense that it tracked with his sense that jobs were the proper route to equality.[44] When Kennedy issued the order, it covered thirty-eight thousand contractors, so its impact was potentially immense.[45] As for the federal workforce itself, the Defense Department, for example, which had desegregated in 1948, had 110,000 African American employees and at the time of Kennedy's executive order and was "probably the biggest employer of Negroes in the world" according to *U.S. News and World Report*.[46] Imagine the impact—the thinking went—if the whole government could hire in a completely race-blind fashion. That is what the PCEEO set out to do.

The idea was that the head of a company could not afford to go to his shareholders and announce the loss of a multi-million-dollar NASA contract because of race-based hiring. A lot of the business community in the South realized desegregation was coming. The space program in those communities gave business and civic leaders an opportunity to accelerate the process and blame the space program, saying, in effect, "Well, we would like to stay true to the old ways, but we just can't. We have these contracts; we have this base, we have this mission."

Taking Affirmative Action

History has long forgotten the PCEEO; however, the executive order creating it has an enduring legacy. The order mandated that federal contractors "take affirmative action to ensure that applicants are employed, and the employees are treated during employment, without regard to race, creed, color, or national origin."[47] Over the years, the term "affirmative action" would come to be synonymous with quotas, especially when it was applied to universities in the late 1970s, but what it represented in 1961 was a shift in government policy on racial discrimination. No longer would the government simply remedy discrimination cases. Instead, it would affirmatively promote and enforce the idea that employers bore a responsibility to hire people of all races. Companies that were equal opportunity employers had to announce that anyone could work for them and then they had to create a work environment where that was true.[48]

The terms "equal opportunity" and "affirmative action" are so commonplace today that it is difficult to understand their impact at the time the Kennedy administration first imposed and enforced them. Frank Crossley understood. He had been working in the IIT Metals Research Department for a few years when he learned about two management openings. Crossley interviewed for both jobs and both supervisors told him that he was their choice. Though he waited and waited, however, no job offers in metals research ever came. Crossley checked into the situation and learned that "at the executive level, someone had turned down my employment." The person who turned him down never interviewed him, so he did not know whether Frank Crossley was qualified. All he knew was that Frank Crossley was black.

Crossley's mentor at the time, John T. Rettaliata, who would go on to be president of IIT, was a refugee from fascism in Italy. He knew discrimination firsthand and, to him, this treatment seemed neither right nor fair. So Rettaliata got on the phone and called the head of the IIT research institute while Crossley listened in on Rettaliata's half of the conversation. As Crossley told it, "He told him that he had heard that Frank Crossley had been offered a job there and had been turned down at the executive level by someone who had not even interviewed him." Crossley said there was a pause and then he heard Dr. Rettaliata say, "No! You cannot do that. [Pause] Why, well because we are an equal opportunity employer." Crossley said when he heard those words he jumped up with a start. "That was the first time I had ever heard the expression 'equal opportunity employer.'" Not too long after the call, the director who had turned him down did interview him, "and I was offered a job after that."[49] That is what "equal opportunity" and "affirmative action" meant in the real world in 1961.

Human Resources

As with Julius Montgomery at Brevard Engineering College, Frank Crossley's story of discrimination demonstrates a different approach from the common story in civil rights histories. Once again, his was a quiet approach. He did not picket or march. He did not organize and he did not sue. While Julius Montgomery's story can be seen as an old school reaction—backing down, trusting a white man in charge, and hoping for the best—Crossley's story demonstrates what was at that time an emerging approach. His integration had the assistance of a set of business management techniques that came to be widely accepted during the war. The period when Crossley was entering the workforce was the heyday of the new field of "human resources." During the war, researchers—many of them with roots in race relations and many of them from Chicago, where he was working—fanned out to factories and offices around the country to teach bosses and managers new ways of working with their employees. Managers were encouraged to understand the hopes, needs, and fears of their workers and to use that understanding, rather than coercion, to get what they wanted. This was also the era of Roosevelt's Fair Employment Practices Committee, which the government charged with integrating industries vital to the war effort.[50] Pamphlets like the 1942 "How Management Can Integrate Negroes in the War Industries" denounced workplace discrimination as just one more way managers and owners were besieged by problematic workers.[51] The war effort gave the consultants making this argument a strong tool. Race discrimination was depicted as un-American—one more impediment to winning the war.[52]

These ideas had taken hold by the time Crossley was rising at IIT. The training he received as an aspiring manager, as well as the predominant workplace culture, had drummed these principles into him and every other up-and-comer after the war.[53] He absorbed these lessons and adapted them to those he had learned back home growing up on the South Side of Chicago—control your emotions, especially anger, and do not give anyone an excuse to make you a stereotype. In the case of the IIT manager's discrimination, Crossley saw injustice and did the thing his management training classes taught him was the right thing. He went to a supervisor. Fortunately enough, the supervisor understood the new rules, too, when it came to integration.

Implementing Equal Opportunity

The civilian space program began a massive hiring binge following Kennedy's man on the Moon speech, so NASA contractors were early participants in the equal opportunity game. Soon after the president's speech, national firms with connections to the Marshall Space Flight Center (MSFC) in Alabama began to advertise for engineers and technicians around the country. Ads run in the fall of 1961 by G.E. and Thiokol Chemical in the *Houston Post* proclaimed the companies to be equal opportunity employers.[54] The advertising requirement was a new element of the Kennedy administration's equal opportunity strategy and the space program made it a large one.

News stories at the time focused on a handful of other elements that made Kennedy's PCEEO seem different. They mentioned that membership on the committee included agency directors (including James Webb, the new administrator of NASA), that it had the ability to investigate and enforce equal hiring by labor unions doing government work, and that it would allow the attorney general to bring legal action if needed.[55] The president was keen to demonstrate that the PCEEO was not going to be "another futile gesture in the chain of Washington attempts to deal with the massive problem of racial discrimination through feckless committees."[56] So almost immediately following the first (ceremonial) meeting, the committee members got together with the leaders of major labor unions and defense contractors.[57]

On the day the president unveiled the PCEEO, its new chairman, Vice President Johnson, declared that when it came to enforcing equality in hiring by federal agencies and contractors, "we mean business."[58] A few days later, NASA administrator Webb wrote to assure the vice president that "I am trying to make it clear in the space administration that you and I together mean business."[59] Webb had his work cut out for him. Knowing as he did that his agency had the worst record in the federal government of hiring African Americans, he called for "the support of all NASA personnel in vigorous action to implement the President's program for equal employment opportunity regardless of race, color, creed or national origin." He designated compliance officers at all the NASA centers.[60] He also sent a memo to the program directors and staff officers of all NASA field installations telling them to "take the initiative to see that the provisions of the Executive Order [creating the PCEEO] are carried out in your area of responsibility."[61] There is evidence that the entirety of this initial message on racial integration did not get through. Tellingly, in the MSFC newsletter, the *Marshall Star*, the May 10 article announcing the new

EEO compliance officer made no mention of race. It said the new officer will "assist the Director in assuring that all civil service and contractor employees directly or indirectly connected with the center have
equal opportunities for employment assignment and advancement."[62]
Nowhere did it say what the words "equal opportunities" referred to,
and one cannot assume that in 1961 people knew offhand what that
term meant. At the time, there were forty-three African Americans
employed at the center.[63]

Planning Progress

In the hours before Kennedy set up the PCEEO, the Pentagon announced a ten-year program to build and procure the C-141 Starlifter,
a massive jet built by the company Frank Crossley had recently joined,
Lockheed. The plane carried equipment and paratroopers to a combat zone. One would also end up at NASA as a flying observatory. This
was the first time the nation had ever spent a billion dollars for a single
military system.[64] What black employment there was in the aerospace
industry in 1961 had been slowing to a trickle in the years since the end
of World War II as companies shut down plants within city limits in
places like Los Angeles and moved out to suburbs like Marietta, Georgia (where the C-141 would be built). Suburban locations were often
beyond the reach of the bus and transit systems used by most African
Americans. In addition, the handful of black workers at the Lockheed-
Marietta plant were ghettoized into an all-black affiliate of the International Association of Machinists, and the white union local denied
these workers access to its apprenticeship program.

This behavior was not unique to Lockheed. In fact, it was typical. There appear to have been few industries more openly contemptu-
ous toward the idea of racial equality at this time than the aerospace
industry. "You will find almost universal prejudice against Negroes"
at aircraft plants, *Fortune* magazine declared in 1943. "There is little
concealment about the anti-Negro policy."[65] Frank Crossley's thoughts
about having to move to Canada or Mexico for work came from an
understanding of this situation. He no doubt knew of the outrageously
racist statements made by aircraft manufacturers. Reporters asked
J. H. "Dutch" Kindleberger, the president and general manager of
North American Aviation (which would go on to build the Apollo space

capsule), about hiring blacks when the company opened its new plant in Kansas City in 1941. He said, "Regardless of their training as aircraft workers, we will not employ them in the North American plant. While we are in complete sympathy with the Negro, it is against the company policy to employ them as mechanics or aircraft workers." African Americans, he said, "would only be hired as janitors."[66] That year the Associated Negro Press wire service carried a story that Boeing Aircraft had "no place for Negro workers" in its plant in Chicago.[67] The year before, W. G. (Gerald) Tuttle, the director of industrial relations at Vultee Aircraft, told the National Negro Congress that "it is not the policy of this company to employ people other than of the Caucasian race."[68] Thus it is of little surprise that of the 250,000 people working in aerospace between 1945 and 1950, African Americans represented only 1.6 percent. That number would improve slightly in the 1950s and then significantly in the early 1960s as federal rules required NASA contractors to hire more African Americans.[69]

The NAACP took advantage of Kennedy's announcement of the executive order creating the PCEEO by challenging discrimination in the aerospace industry head-on—filing thirty-two complaints on behalf of black Lockheed-Marietta employees.[70] Here on the very first day of the PCEEO's existence was an immediate challenge to the Kennedy administration's approach to equal rights through federal contracting. On its commitment to end flagrant discrimination and smooth the path to southern black employment in federally supported defense plants, would the administration put up or shut up?

PCEEO executive director John Feild flew out to Lockheed corporate headquarters in California to talk with Courlandt Gross, the company's president. When they were through, Gross had agreed to two actions, neither of which addressed the problem at hand. First, Lockheed removed the "white" and "colored" signs from the bathrooms, water fountains, and cafeterias in the Marietta plant. Second, on May 25, in a high-profile White House ceremony, Gross signed what was termed a "Plan for Progress" to eliminate segregation at the facility.[71] In a single photo op, Lockheed earned itself a reprieve (the NAACP had secured a pledge to cancel the contract if the company did not comply with the president's EEO order)[72] and "the reputation as the region's most active and interested employer of Negroes."[73] The Kennedy administration bought itself some effective public relations, and

Lyndon Johnson was able to brag that he really did "mean business" when it came to workplace equality. The model was a Washington classic and before you could say "too good to be true," defense contractors were lining up at the White House gate for a chance to shake the president's hand and sign Plans for Progress of their own. On July 12, eight more defense contractors showed up. On November 24, twelve more. On February 7, 1962, thirty-one came by.[74] In all, by February 1964, "one hundred forty one corporations employing approximately 7 million workers [had] signed Plans for Progress with the federal government."[75]

From one perspective, this was the Kennedy administration taking bold action to guarantee African Americans a slice of the federal pie. The NAACP, however, did not view it from that perspective. Neither did the National Urban League. Nor the black press. The National Urban League pointed out that none of the Plans for Progress had compliance procedures or addressed any of the things that companies were required to do under the president's executive order. The National Urban League further noted that the program had a loophole allowing companies to claim they were complying with the spirit of equal employment opportunity while actually not doing anything at all. At its national convention in July, the NAACP passed a resolution calling Plans for Progress "virtually useless." One analysis showed that in those 141 companies employing 7 million people, Plans for Progress had resulted in only 2,000 African Americans getting jobs.[76]

Plans for Progress—which in Vice President Johnson's mind became synonymous with the PCEEO—constituted only one link in the system, and a weak one at that.[77] As Johnson wrote in the executive order, contractors were not required to police their subcontractors, and they did not have to guarantee subcontractor compliance, either. All they had to do was insert an equal opportunity clause in their subcontracts and require that the subcontractors file compliance reports.[78] The compliance system set up under the executive order had its own enormous loophole. Contractors had to submit progress reports and they had to open their books to the committee, "but this requirement could be waived if the committee awarded the contractor a 'certificate of merit.'"[79] Though Johnson went so far as to call cabinet secretaries in the middle of the night to quiz them on their EEO compliance, he complained bitterly that the structure of the PCEEO left him insufficient resources to carry out the committee's mandate.[80] In fact, there are hints that the PCEEO was intentionally set up this way, and for a specific reason: under its structure, the Kennedy brothers got credit for any good the committee did, and when things went wrong, Johnson got the blame.[81]

Regardless of the weakness of its structure and the questions raised about the implications of its creation and the insufficiency of its procedures by the national civil rights leadership and the black press, the PCEEO was the structure that the Kennedy administration chose to implement Johnson's plan to create equality in the South through jobs. Flawed as it was, the PCEEO would be the instrument they used to drive NASA and the space program to examine and modify the way they did business in the early 1960s. The task was huge, the tool chosen insufficient. Given the political realities, however, it was the president's choice; and, over time, Johnson's experiment—creating an entirely new class of jobs for African Americans to aspire to—would attain a measure of success. The number of employed black professional and skilled laborers in Houston increased during the 1960s.[82] And according Herbert Northrup of the Wharton School of Business, once "civilian aircraft production, military procurement, and the space program all boomed in concert with increased emphasis on civil rights . . . Negro employment in the aerospace industry began to show both qualitative and quantitative advances."[83] This new committee, with its power to cajole, investigate, and possibly force integration, would turn out to be one reason why.

"No matter how dark the future looks, the young Negro student must prepare himself well," Crossley told S. I. Hayakawa back in 1945. "Opportunities for Negroes are not great, but they are gradually increasing. You cannot take advantage of an opportunity unless you are prepared."[84] Crossley certainly did that. He was the first African American to receive a PhD in metallurgical engineering.[85] He received seven patents, five for the titanium-base alloys that greatly improved the aircraft and aerospace industry. He received something more, as well: the respect of his coworkers. Many years later, it is difficult to recall how monumentally important that was. When Crossley began his career, race discrimination was an accepted fact of everyday life. A coworker could walk up to someone like Julius Montgomery and declare, "You are nothing but a Nigger," with impunity. In the Hayakawa interview, Crossley was referring to this dark future—the one he set out to change.

Crossley did not know it in 1945, but he was preparing himself for the Space Age, which would soon arrive. He was able to take advantage of it. The PCEEO and its members would push NASA to see that other African Americans could, too. Despite a massive effort, they would largely fail. As we will see, the critics of America's inactivity on racial integration took their opportunities, too, and when those opportunities included outer space, the results could be memorable.

Dixie's Role in the Space Age

It is imperative that we depart from timeworn traditions and concepts and adopt space-age techniques to cope with the problems of our space-age cities.

WAYNE THOMPSON, CITY MANAGER, OAKLAND, CA

The kind of optimism S. I. Hayakawa expressed was commonplace as World War II ended. The defeat of racist totalitarianism in Europe and Asia—achieved largely through American scientific and technological superiority—left American opinion leaders and the public at large with a sense that they were living in a very special place at the cusp of a very special time. That sense of America's eminence was cultivated and enhanced in the years between the war's end and the advent of what is now called the "Space Age." During the Truman years, elites in and out of government created and disseminated "legitimating ideologies" designed to bring the nation in line with the administration's international grand strategy.[1] As the 1950s progressed, the suppression of "un-American" ideas helped enforce those ideologies.[2] By 1961, when President Kennedy was challenging the nation to put a man on the Moon, these ideas had coalesced into a "pervasive American consensus," one that can be summed up as, "the United States is better and different."[3]

In light of this national ethos, it is easy to understand why there are those who remember the Space Age and pine for its return with a deep nostalgia—especially considering what happened to that sense of American harmony following the turmoil of Vietnam, civil rights, Watergate, and everything that followed. By the time the Apollo program ended in 1972, America was well on its way to a new consensus about itself—one that "was critical of older interpretations [and] easy generalizations" and much more difficult to love and embrace blindly.[4] The Kennedy years represent the twilight of the postwar ethos, when it was up to the United States to set things right—a type of thinking that went hand in hand with what was portrayed as the "awesome adventure and . . . noble challenge" of space.[5]

The allure of space—and space travel in particular—thoroughly seized American popular culture. For a while, it seemed that everything was about space or at least made reference to it. All three of the television networks carried the launches, near launches, and scrubbed launches of NASA rockets, and it was not long before the space program spun from engineering to entertainment and from geopolitics to product. America fell in love with a government agency that spent billions of dollars launching rockets and sending men into space; at the time, it was easy to ask, how could we not? America loves rockets. Our national anthem sings of rockets. We are also a nation committed to building bigger, better, and faster machines. Nothing was bigger than a Saturn V rocket. Moreover, the message went, Americans are restless people who are always seeking new frontiers. This mentality joined with postwar optimism, wealth, and the rapid growth of aerospace technology to turn the science fiction of H. G. Wells and Jules Verne into reality. Vice principals rolled TVs into classrooms to watch lift-offs. Teachers let kids listen to a radio or watch a portable TV during class when a capsule was due to splash down. Teachers filled bulletin boards with space drawings, space newspaper clippings, and *National Geographic* space maps, while young and old filled bookshelves at home and school with new space books, some of which even had photographs!

The space program became an entertainment theme in its own right. Songs, movies, comedians, and television programs all found a way to connect themselves to NASA. "Telestar," by The Tornadoes, arguably the best pop song ever written about a communications satellite, became a No. 1 single in the United States in 1962—a first for any British band. There was 1968's "The Walk of Ed White" by Up with People, about the first spacewalk, during Gemini 4. In 1969, The Byrds' *Ballad of Easy Rider* included a short track called "Armstrong, Aldrin, and Collins." Movie studios and television networks produced space, Space Age, and space exploration fare throughout the 1960s.

In *Hold On! (There's No Place Like Space)*, NASA names a rocket after Herman's Hermits, a British pop band, while they are touring the United States in 1966. Also in 1966 was Jerry Lewis's *Way . . . Way Out*, in which the Soviets and Americans send male-female teams to the Moon. Advertising copy announced, "The Space race gets Racier when . . . American and Russian Astro-Nuts and Commie-Cuties mix it up to see who will be the first to populate the moon!" Don Knotts starred in *The Reluctant Astronaut*, where he played a rocket-ride operator at a city park whose father applies to NASA on his behalf.[6] Among the better-known TV series of the era was *The Jetsons* (ABC, 1962–1963), a primetime cartoon that showed the futuristic life of George Jetson, his boy Elroy, daughter Judy, Jane, his wife, and Astro the dog. Other shows included *Lost in Space* (CBS, 1965–1968), about a family marooned on an unknown planet, and *Star Trek* (NBC, 1966–1969), which followed the USS *Enterprise* as it went boldly "where no man has gone before." There was also one show completely about NASA (sort of) that managed to survive into the 1970s. That was *I Dream of Jeannie* on NBC. The plot was simple: astronaut Anthony Nelson and his space capsule land on a small desert island. He finds and opens a bottle, which releases a beautiful blonde genie named Jeannie who immediately falls in love with her new master.[7]

Space exploration and NASA influenced American culture well beyond books, movies, and television. When NASA arrived in Houston, the town's baseball team, the Colt .45s, changed their name to the Astros. Of course, the team renamed their new indoor stadium the Astrodome. Ticket girls, called "Spacettes" and dressed in space-themed uniforms, helped people to their seats. In 1971, a professional basketball team, the Rockets, arrived. In other communities, building the Space Age was less dramatic and cheaper. Cities throughout the United States added Astro-City playgrounds to municipal parks. The centerpiece of this complex was the rocket slide—a spiral staircase inside of a steel cage with a steel slide extending from the side. It was fun, scary, and dangerous. Just like a real space rocket.

Deus ex Machina

The unifying emphasis behind this cultural amalgam was the idea of The Future. Humans were primed to land on a foreign planet for the first time in history. That recognition wrapped the Space Age in an envelope of possibility that extended beyond rockets to permeate every aspect of life, including race relations. Understanding how this came to be requires a thorough grasp of the sociological import of space imagery and space iconography. In its time, the term "Space Age" meant so much more than advances in engineering. This new age of space was for many a panacea—a deus ex machina. Those who invoked it thought it could solve almost any problem. In the early 1960s, many Americans saw the earth as grubby and brown, possibly dying, and certainly choked by air and water pollution.[8] Cities were beset by hideous architecture and crumbling infrastructure, while countries warred needlessly over ideology and resources.[9] Contrast that with outer space. Outer space seemed clean and peaceful. The powerful and beautifully designed machines that took humans there were brand new. The Space Age offered unending horizons.

The boundless optimism of this period was on stark display, for example, in the proceedings of a NASA conference held in Oakland, California, in the early 1960s. The conference's subject matter—space, science, and urban life—guaranteed that the wishful thinking would be set in particularly high relief as the Space Age was placed hard up against the contemporary vision of America's roiling urban landscape. Owing perhaps to NASA's orientation as a technology agency, the conference tended to deal with tools for improving city services (sanitation, highways, and water quality) rather than with actual city dwellers. But because it was a conference on designing the city of the future, it offers a particularly revealing opportunity to see in real time what the term "Space Age" meant to the people who used it. A striking example came in the welcoming remarks given by Oakland city manager Wayne Thompson. He began predictably enough, using "Space Age" as a handy piece of jargon. "It is imperative," Thompson said, "that we depart from timeworn traditions and concepts and adopt space-age techniques to cope with the problems of our space-age cities."[10] But then he went on to delineate what some of those "space-age techniques" might be, and here he actually opens a window of meaning to a modern-day reader. What was a "Space-age technique"? Well, someday, he said, a police officer preparing for a traffic stop could "read the license number of the car he is following into the central police databank, and the computer will within seconds return to him the name of the driver, whether he has a police record, where he lives, and so forth." Space Age techniques for the home would include "a telephone that will automatically dial the police department when a burglar enters a residence" and "a fire detection device that will automatically dial the fire department when a fire occurs." He also talked about solar and nuclear energy, as well as high-speed travel that "will make close neighbors out of great metropolitan centers like Washington, DC, and Oakland."[11] Space Aged in the early 1960s meant the same thing as "modern" in the 1920s, "high-tech" in the 1980s, "cyber" in the 1990s, or "cutting-edge" in the early years of the twenty-first century. As with those other words, a thing was "Space Aged" if it would solve a problem of today tomorrow. It did not matter what the problem was.

Shortly after troops ended rioting over the integration of the University of Mississippi, Governor Ross Barnett attended a southern governors' conference titled Dixie's Role in the Space Age, where he led a session called "Space Age Problems."[12] During the session, Kentucky governor Bert Combs chastised Barnett, calling on him to "quit playing Custer's Last Stand and join us. We need to talk about freight rates and other Space Age problems."[13] Freight rates in the Space Age? Yes, and not only that.

Speaking in Seattle in 1962, Vice President Johnson made a speech entitled "The New World of Space." In it, he demonstrated his whole-hearted embrace of the idea that the Space Age could solve anything. "Because the Space Age is here," Johnson said, "we are recruiting the best talent regardless of race or religion, and, more importantly, senseless patterns of discrimination in employment are being broken up."[14] What Johnson said is standard fare for the early 1960s. But take a closer look. The elements seem almost completely unrelated. "Senseless patterns of discrimination in employment" are being broken up because "the Space Age is here"? What do those two things even have to do with each other? In the narrow sense, nothing: a country does not end race discrimination by going into outer space and putting an astronaut on the Moon. But the quote does make sense within the paradigm of a time when many a sentence began with "Because the Space Age is here . . ."

The vice president was not the only one tying the space program to race relations. While the eyes of the nation were on NASA in its earliest days (1957–1961), it is nearly impossible to find a newspaper article in the mainstream press written about how the agency was doing on race relations. However, while the *Huntsville Times, Houston Post, Melbourne Times, Titusville Star-Advocate*, or even the *New York Times*, for that matter, did not touch the issue, there was another set of newspapers that did. Those were the papers known as the black press. Many people in America caught a dose of space fever as Kennedy launched the nation into the space race, and the black press did its part to rev up its readers and get them excited about this challenging new human endeavor, running wire service stories with headlines like "Astronaut Grissom Rarin' to Go."[15] When it came to black papers expressing their own opinions on NASA, however, the tone and the content were often quite different.

Space Age Ideals

James Hicks, the executive editor of the *New York Amsterdam News*, for one, was a man with a constant eye on the state of racial progress who never feared speaking his mind when he saw backsliding or hypocrisy. In a scathing column that ran after Yuri Gagarin became the first human being in space, Hicks pointed out that "as the Russian engineers made mankind's greatest history there was not a single Negro engineering student attending an accredited engineering school south of Washington, DC, for the simple reason that they are shut out because of their race." He also used the space shot as an excuse to ask readers to note that "in New Orleans, Louisiana, white people who should have been helping America in the race with Russia, were busy ganging up on and attacking four little Negro girls whose parents dared to send them to a white school built with the aid of US government funds." He summed up by saying, "If the United States of America had spent as much time PUSHING AHEAD in the race for space with Russia as it did HOLDING BACK the Negro BECAUSE of his race—America, and not Russia, would be accepting the congratulations of the world today."[16] Hicks' column demonstrated a style of critique that the black press deployed across the country—placing NASA's image as a harbinger of the future against the backward racial policies of the communities where the agency had its centers. NASA's image was clean, new, and untainted by the problems in the South. The black press, however, liked to get beyond the image and show some of the rot beneath the façade.

An elaborate example of the black press using Space Age rhetoric is an article that ran in the *Atlanta Daily World* on May 25, 1962, which when read decades later may seem simply bizarre, but which makes sense in the context of 1962. "The space age world," the article begins, "was a mixture of pride, anxiety, suspicion and disappointment Thursday as Astronaut M. Scott Carpenter had his troubles in the Atlantic Ocean, while Negroes in a town not too far away saw their apparently successful plan of voting a colored man into public office, go up in smoke." The previous day, Carpenter was temporarily lost at sea after his space capsule overshot its splashdown point, which explains the article's reference to "anxiety." But what was the other event that caused "the space age world" suspicion and disappointment? It turns out that an African American insurance agent named Earl Baggs had lost a close election for county commissioner in Hinesville, Georgia, to a white man named Hoke Youmans when someone disqualified three hundred black votes under suspicious circumstances. What did this have to do with space? As in Vice President Johnson's speech—nothing, in the narrow sense. Nevertheless, that did not stop the newspaper from tying the election to the space program.

"A brave young man had blasted off into space in a new ship, showing we were on the ball in going modern with rockets. But here on earth other brave people, who had dared to use Space Age ideals against prejudiced tradition, felt they were still hard up against an old boat of inequality." Space Age ideals. If Baggs had won, the paper said, "for the Negro in Dixie it would be as big news as any astronaut orbiting the Earth successfully." It was 1962, the Space Age, but "once more a Negro candidate had been defeated in the Southland, which called for beating Russia in space, while stubbornly holding on to past traditions, despite efforts by the last three presidents to show that equality is needed to bring success to space programs and every other ideal in this country."[17]

Here again was the black press placing NASA—symbol of the "New Frontier" and the brightest jewel in the New South's economic and industrial "second Reconstruction"—against the reality of southern intransigence toward racial integration. Second and more basically, however, this was just another example of people seeing the Space Age as the thing that would solve today's problems tomorrow. Whether it was southern governors discussing Dixie's role in Space Age problems or a black southern newspaper looking at interposition and nullification, the Space Age was conveniently present to address the problem at hand. "Here the southern Negro watched Carpenter take off into space Thursday morning with a heavy heart which had been forced once more to realize that this was after all still Dixie."[18]

In Houston, We Have a Problem

All of this forms the context in which Otis King was operating when he set out to change the archaic state of race relations in a city perceived to be so thoroughly forward thinking that its name would become the first word ever spoken by a human being on another planet.[19] Houston, as one of the largest cities in the United States, had a level of economic diversity that allowed the number of employed black professional and skilled laborers to increase during the 1960s. The Houston Chapter of the Episcopal Society for Cultural and Racial Unity reported in 1964 that, due in part to government pressure, employers were "seeking qualified Negroes for responsible positions."[20] *Ebony* magazine even named Houston one of the ten best cities for negro employment based upon the rate of job growth, the unemployment rate, and the opportunity for blacks to compete successfully in the job market.[21] Nevertheless, as Otis King understood, Houston was still the South.

King was not a pioneer of the space program. He worked as a lawyer, a professor, and a civil rights activist his entire life. It still makes sense to profile him here because of his unique legacy. King was responsible for perhaps the most effective use of NASA's Space Aged imagery in the cause of racial integration.

King's awareness of Houston's race problems began when he was a child. "When my mother and I would go to town on the bus," he recalled, "we had to be careful to notice where the designation line was placed for 'colored' and make sure we sat behind that. It was a moveable sign. I think they could clip it onto the seat, because, at times, when the bus population changed and there were more whites, they would move it farther back so that the whites could still sit in front of blacks." Regardless of what the Houston Chamber of Commerce or the nation at large thought, King said, "Growing up in Houston my feeling—and I believe the feeling of my community—is that we were in the South, just like everyone else in the South. The notion that Texas was West or Southwest came along much later." In NASA's most iconic home city, he said, "we were totally segregated and restricted in terms of movement. There was no difference in my thinking and my feeling about living in Houston than . . . about persons who lived in Louisiana, Mississippi, or anywhere else in the South."[22]

For King, the desire to demand change came about at an age when many express outrage, during the teenage years. "My motivation for protesting segregation started quite a while before I actively participated in the movement," he said. "It started as a kind of recognition that some things just were out of balance. As a young child, it was difficult to understand why me and my mother had to sit behind a particular line. As an older child it was difficult to understand why I could not get a drink of water or use a restroom in certain facilities downtown." These indignities, he said, "kind of fester, and when the opportunities became available, to speak out—to participate and do something about it—I jumped at it. In other words, I was kind of ready; just waiting for it to come about."

Few would include Houston as one of America's hotbeds of race discrimination in the 1960s. The fact is, though, the city was where the Texas Ku Klux Klan began. It also had the largest segregated school system in America for most of the twentieth century.[23] By 1963, King said, Houston viewed itself "as a progressive city," but he said the truth was otherwise and that racial segregation was "totally pervasive."[24] As a 1997 *Houston Chronicle* article commemorating the end of the city's segregation put it, "If you were black you couldn't eat at a white restaurant, except maybe around back. You couldn't use the restroom in a white establishment. A white ambulance wouldn't pick you up if you lay bleeding on the street." Otis King's recollections were reflected in the article's summation that "in none of these respects was Houston any different from other cities in the South."[25] According to Quentin Mease, director of Houston's black YMCA, an African American could not eat lunch in the cafeterias at city hall or the county courthouse. The city's hotels were "absolutely segregated," and, he said, when professional football came to Houston in 1960, the only place Oilers management allowed blacks to sit was "behind the goalposts."[26] Overall, Mease said, "unequal rights, denial of opportunities, [and a] lack of freedom of access" were all part of daily life for African Americans in Houston.[27]

As in Florida and Alabama, these racist policies and laws clashed with the federal government's efforts on racial equality and caused headaches for the largest federal entity in the area, NASA's space centers. In 1961, Houston Power and Light cut power to the Pelican Island Destroyer Base in Galveston because the company objected to the inclusion of an antidiscrimination clause in its contract with the navy. At the time, Houston was on a short list of places that might house NASA's Manned Spacecraft Center (MSC). Vice President Johnson called Houston congressman Albert Thomas and, demonstrating a type of flair that made him legendary, asked, "Shall I tell the President that you cannot supply power to a Navy installation there because of the Negro question? And . . . what are you gonna do about space?" The navy got its power. Many also see the phone call as allowing Houston to remain a viable site for NASA's Manned Spacecraft Center.[28]

The Free Instruction of White Texans

A much larger problem came up a year later and threatened to derail Houston's status as a space community entirely. The problem centered on the Rice Institute, a private institution of higher education that changed its name to Rice University in July 1960. The school president, Kenneth Pitzer, was a point man for Houston during the MSC site selection process. He took that opportunity to confer with NASA administrator James Webb and members of the site selection committee whenever the school called on him—as it often did—to play tour guide during NASA visits to town.[29] Though Houston and Mission Control are now synonymous, in the early years of the Kennedy administration, the site selection decision for the MSC was up for grabs and there were many places in America that NASA might have chosen. An August 24, 1961, headline in the *Houston Chronicle* confidently declared, "Chances Are 99–1 That Houston Will Get the Proposed Space Lab."[30] During this time, however, Humble Oil gave Rice a thousand acres near Clear Lake, Texas, and on one of those site visits, Pitzer promised the land to NASA for the MSC building. At that time, the land was valued at $3,500 per acre—a gift of $3.5 million. According to Albert Thomas, a Rice alumnus, the deal "tipped the scales" toward NASA's selection of Clear Lake.[31] Rice did not just give to NASA; it also received. By 1962, the school had gotten more money in space research grants than Texas A&M University and the University of Houston combined. It also sought to establish the first Department of Space Science in the United States.[32] Therein lay the problem.

William Marsh Rice founded the Rice Institute in 1891 to provide for the free instruction of *white* Texans. Although the institute admitted Asian students after World War II, by the 1960s, "there were still no black students on campus," and the federal government could deny a segregated Rice federal money. That included NASA dollars.[33] This was a problem Rice and NASA had been stumbling their way through since the space agency's creation. The Kennedy administration's emphasis on space and race brought the issue to a head, and the college's land grant to NASA meant it was going to have to take care of it once and for all.[34]

Pitzer was not from the South and he found the racial wording in the school's charter "just ridiculous."[35] George Brown and J. Newton Rayzor, the two Rice trustees who had most actively recruited Pitzer, shared that opinion. Apart from those two, however, there was no real appetite for desegregation among the college's trustees, no eagerness to step forward and make what they believed would be a very unpopular change.

At a board meeting in May 1962, when the issue was first raised, "no one, apparently, contended that it was wrong or offensive or in violation of southern tradition to allow blacks to enroll at Rice. Instead, several trustees argued that taking steps to remove the bar on black admissions would actually hinder rather than help Rice's efforts to solve its financial problem."[36] A 1961 referendum showed Rice's undergraduates favored integration by a "two-to-one margin, the graduate students by four-to-one, and the faculty by eight-to-one," but many alumni felt the same ambivalence as the board of trustees. Brown and Rayzor extracted a statement from the alumni association saying that "due to these changing times, we should set our personal feelings to one side" and consider allowing African Americans to attend. But that did not settle the issue and the fight on the board continued. On September 26, four months after the issue first came up (at the same meeting where Rice agreed to give the land in Clear Lake to NASA), trustees unanimously approved a motion made by Newton Rayzor to file "a lawsuit to change Rice's charter . . . and admit black students."[37] Several more months passed before, in February 1963, the university's lawyers filed suit to remove the words "white" and "free" from the founder's indenture.

Many alumni misunderstood the suit's purpose, "even the majority who approved of the action." They "understood the lawsuit to be a challenge to William Marsh Rice's will, not a rather ordinary petition to change the interpretation of trust language."[38] In a telegram, "two enraged alumni, [who] could not wait for the postal service to send their disapproval to the university, said that they were 'highly disappointed and amazed at incredible action of board seeking to destroy basic and crystal clear provisions of Rice will. Trust that court will not uphold this flagrant and arrogant violation of our benefactor's desires.'"[39] Two others, John Coffee and Val Billups, went even further to express their disdain for integration. They filed an intervention plea insisting on strict interpretation of the indenture. As a result, Rice's relationship to NASA and NASA's ability to occupy the land where it hoped to build the MSC hung in the balance until 1965, when a jury finally ruled that restrictions on race and tuition were impractical and hindered Rice University's development.[40]

The absolute number of black professional and skilled technicians at NASA in Houston was, overall, abysmal, and Otis King said that was no accident. As if it were a corollary to William Marsh Rice's racist instructions regarding admission, the site selected for Mission Control practically guaranteed that NASA would only be available for the employment of white Texans. The Clear Lake area between Houston and Galveston, King said, had "several small towns. Our thinking during that time was that they were basically sort of Ku Klux Klan territory." In other words, he said, "You had a stretch of about fifty miles where there were blacks in Houston and then blacks again in Galveston, but no blacks living in between, and NASA of course was right in the middle of that in space."[41]

African Americans in the South were much less likely than whites to have their own cars in the early 1960s, so the problem of distance was severe.[42] "Anyone who worked at NASA," King said, "either had to drive from Houston, which was twenty-five or thirty miles, or drive from Galveston, which was twenty-five or thirty miles. There were no blacks living in the surrounding area." There was housing around Cape Canaveral, lots of it, but that did not solve the problem either. Between 1963 and 1965, 1,200 new residents had moved to Clear Lake City; by 1968, that number had increased to 50,000.[43] But among the families that poured into the area, not even one was black.[44] There were only sixteen African American students at Clear Creek High School in 1968 out of a total student population of 2,391.[45] Here was the space program, promoted by its adherents as confirmation that America was a country "better and different," clearly demonstrating that it was not.[46] Otis King knew that and he knew it was wrong.

Space Brings Us Together

Debate flared throughout the 1960s over just what, exactly, America was getting for the money it was spending on space exploration. Although news coverage of the space program often cloaked NASA's efforts in the vestments of science, it was often a vague science—nothing that held immediate benefits but instead, as George Simpson, a NASA assistant administrator, declared at the conference in Oakland, "great promise for civilian applications in the not too distant future."[47] President Kennedy had his own reasons for sending men into orbit. As he told James Webb in a secretly taped Oval Office conversation, he was "not that interested in space" and believed that "everything that we do ought to really be tied in to getting onto the Moon ahead of the Russians."[48] In the Cold War battle to see whose economic system would dominate the world, persuasion played an enormous part, and there was something extremely persuasive to the undecided nations of the world—to the Indians or the people in Burundi—about seeing the words "United States" written across the 364-foot Saturn V rocket on its way to the Moon.

The U.S. government often exploited the space program in order to reach out to Third World countries when America wanted to make a good impression. In the case of Africa, it did this in ways that simultaneously allowed NASA to make a good impression on African Americans. The March 3, 1962, edition of the *Baltimore Afro-American* contained the story of four black men who tracked John Glenn's flight from a station in Kano, Nigeria. "Cooper Relays 'Hello' to Africa from

Space Craft" was the headline of an article planted in the *Chicago Defender* right at the end of the Birmingham unrest. The U.S. Information Agency (USIA) hired African Americans to work in Africa and they assigned two to give lectures on the space program.[49] When John Glenn's *Friendship 7* capsule went to Africa during a world tour, twelve thousand Ghanaians saw it on its first day in Accra. In total over fifty thousand people viewed *Friendship 7* there, a number that surpassed attendance in London, Paris, Belgrade, and Madrid. The USIA took extraordinary steps as the capsule traveled from Ghana to its next stop in Nigeria, with agency director Edward R. Murrow declaring that "it was imperative to show in the exhibit 'all American Negro personnel working with white staff on project.'" Nearly nineteen thousand came to see the spacecraft during its time on display in Lagos and Kano.[50]

Aside from the planned reasons for having a space program, there turned out to be another unforeseen consequence—one that tended to reinforce a sense of American consensus at precisely those times when it was most necessary. In the words of CBS News anchor Walter Cronkite, "the success of our space program in that terrible decade of the 60s played an important part in maintaining a semblance of morale" in America. "The civil rights fight" had split the country "in a way that we hadn't been split since the Civil War of the 1860s," Cronkite said. These battles, as they became more violent and as they came to coincide with clashes on America's streets over the Vietnam War, left the country "very, very depressed," Cronkite said, but the space program was able to bind up the nation's wounds and reassure people that everything was going to be OK. "Here was this one program where people could look up and dream—if you please—of incredible adventure," he said. Pride in the space program "had a great deal to do with maintaining some sense of balance in this civilization of ours."[51]

This is precisely what happened just hours after a settlement was reached in the terrifying racial unrest in Birmingham, Alabama, when on May 15, 1963, astronaut Gordon Cooper left the earth for a 34½-hour flight—the longest ever taken by an American. As Cooper circled the earth at a maximum speed of 17,546 miles per hour every ninety-three minutes, the regular, real-time updates of his physical condition, his blood pressure and heart rate, transfixed America. Although Cooper's flight "fell far short of the four- and three-day missions flown by two Russians" the previous summer, he was nonetheless treated like a conquering hero on his return to Earth. Two hundred thousand people cheered him during a parade in Washington, DC (where he also addressed a joint session of Congress) and between 4 and 4.5 million showered him with love and three thousand tons of torn paper in New York City.[52]

A Hero's Welcome

While hero worship had diverted attention from the country's racial divisions, a group of civil rights protesters had a plan to turn that process on its head. Three people—two young, one old, led this group. They were Quentin Mease, Eldrewey Stearns, and Otis King. NASA planned yet another parade for Cooper after he left New York, this one in Mission Control's hometown of Houston. Mease, Stearns, and King planned to take astronaut lionization and convert it back into a tool to get the nation to see the subject they cared about—racial discrimination.

Quentin Mease was born in Iowa and lived to be one hundred years old, with most of that time spent fighting for the rights of African Americans. "I had been a member of the NAACP when I was just a youth," he said, and though Iowa, like Houston, would not top anyone's list of civil rights cauldrons, he said life there for African Americans was no joy ride. "There was some segregation in facilities there in Iowa," he said, "so we weren't exactly uninitiated about segregation. We knew what it was and knew how to use means to fight it." He kept those early lessons with him when the opportunity arrived to take over the black YMCA in Houston. "There was a difference all right," he said of segregation in the South compared to what he had experienced in Iowa; but he was ready for it and spent the next seventy-some years as a thorn in the side of the city's white establishment. In fact, Mease is widely credited with making sure that blacks could sit anywhere in the Astrodome when it opened. The Astros' original owners were planning to meet with the head of the National League about getting a franchise; Quentin Mease preempted their visit by dictating a "letter to my secretary that was sent up there and said that we supported the idea of getting National League baseball in Houston, provided that there would be no segregation."[53]

Otis King was a law student at all-black Texas Southern University

(TSU) in 1963, just back from service in Vietnam, where life on post had been completely integrated. By the time he returned home, he said, "I had gone through that kind of transitional period and was determined that I would not, sort of, take it anymore." Eldrewey Stearns, who would later succumb to mental illness, was known for his eccentric behavior. Mease remembered that Stearns showed up at the Y one day while some TSU students were setting up an assembly; he walked on stage and recited the Gettysburg Address. Later, Stearns would spend many years in and out of institutions, a victim of alcoholism and bipolar disorder. But for much of the time during Houston's civil rights battle, he was fit and ready, and many considered him to have led the city's sit-in movement.[54]

At the time of Cooper's Houston parade, thanks in part to Stearns' activism and Mease's negotiations with the white business elite, the city had made considerable progress in opening its facilities to people of all races. A great deal of work remained, however. Houston had integrated its lunch counters and hotels, but the city's principal movie theaters and its restaurants were still off limits to blacks. Stearns, who like King was a law student at TSU, started a group called the Progressive Youth Association (PYA), which began sitting-in at Houston's downtown lunch counters in 1961. Not long after the PYA got started, TSU came under pressure to stop the group. When the university caved in, Stearns, King, and several officers of the PYA walked the two blocks down Wheeler Avenue from the university to the Y to talk to Quentin Mease. "They told me they were ordered off the campus and asked if they could meet at the YMCA," Mease said. When he told them yes, it was the beginning of an important relationship. At its start, the PYA could turn out two or three hundred students for gatherings at the Y's gym. These mass meetings attracted the media (including a young TV reporter from Channel 11 named Dan Rather) and the attention of the police, who, Mease said, liked to park their squad cars in the lot behind the gym and then stand outside as "a form of intimidation."[55]

The media covered violent clashes over race in many American cities, but when desegregation came to Houston's lunch counters, it came in silence, total silence in fact—an absolute news blackout that took activists like Otis King by surprise. "We became aware that the media was not covering all of our activities," he said, "primarily because of people who contacted us from other cities and told us about things that they were seeing in the papers there or on television that we were not seeing here."[56]

The business community and the city's media had formed a pact. Under their plan, seventy downtown Houston lunch counters were simultaneously integrated in August 1961 and "for a week to 10 days, not a word of it would appear in local newspapers (white or black), on television, on the radio." A documentary film, *The Strange Demise of Jim Crow*, credits a public relations man named Bob Dundas with organizing the deal. According to the documentary, Dundas worked with the publisher of the *Houston Chronicle*, John T. Jones, who also owned Channel 13 and its sister radio station; Oveta Culp Hobby, owner of the *Houston Post*; and Jack Harris, the manager of the other big TV and radio station in town. The news blackout was designed to make sure integration happened "without violence, with no ugly scenes," and in a way where "no individual lunch counter operator would be exposing himself to segregationist reprisals."[57]

Although the outcome was positive for African Americans who wanted lunch downtown, it undercut the momentum of the PYA. "Protests or sit-in movements such as ours really live on publicity," Otis King said. Desegregation of restaurants and the movie theaters remained unfinished business and "we wanted to put pressure—bring pressure to bear on the powers-that-be by having [our activities] distributed far and wide." The news blackout, PYA members thought, "was deliberate, and was designed to block the impact of what we were doing." That is where Cooper's parade came in. "We picked the event of the parade because we knew that it was one that would have national news coverage," King said. "We were concerned that the events that we participated in would get wide coverage because we knew that that would increase the impact of what we were doing, so we just saw this as a great opportunity."

Stearns and King cooked up an idea guaranteed to get attention. "What we were planning to do," King said, "was really to block the parade." Early in the morning in the TSU dorms, students carefully folded large paper protest signs and wrapped them around their bodies or stuffed them under their shirts. Then they headed out, walking or boarding the backs of buses headed from the city's Greater Third Ward to Main Street. The PYA had "probably about forty people down, mixed in the crowd," and the plan was that "about half of them would have moved into the street and blocked the parade from the front and about half would move behind and keep them from turning around or going in another direction."

It would have been a serious blow and would have exposed Houston to the thing its power structure wanted least, evidence that the city was not what they said it was, that it did have a race problem, and a serious one. "The chamber of commerce had sort of sent the word," King said, that they wanted to "protect the image of Houston as a city that was moving forward and was more progressive than these other southern cities."[58] King and Stearns' plan was going to dash that image to bits.

They leaked word of their plan to Quentin Mease. Up to this time, Mease had been content to work with the city's business establishment to make slow progress toward solving Houston's racial problems. But, in a split that mirrored the divide in the larger civil rights movement, the younger crowd, led by the PYA chiefs King and Stearns (whom Mease called "a feisty little fella"), were impatient.[59] Segregation had to end. Moreover, the blackout around integration had to end. "They were doing it deliberately to try to block the effect of the movement," King said, and he had had it. The old folks could not dawdle anymore, Stearns and King felt. The time for action was now.

When Stearns and King spoke to Mease, they kept the target vague. "We didn't tell him exactly what we were going to do, but we did tell him it was going to make the national news because of the nature of this particular event." Meanwhile, Mease had recruited the PCEEO's Hobart Taylor to talk to John T. Jones about supporting desegregation. In addition to owning the *Chronicle* and the TV and radio station, Jones and his family owned properties "leased to the downtown theaters."[60] Taylor had engaged Jones in a conversation with the management of Loews Theaters in New York, but progress had been typically slow and "by May 22, the day before the parade, the students and their elders had received no assurances an understanding had been reached."[61]

"We Had a Failsafe Point"

On the day of the parade, unbeknownst to Cooper, NASA, or parade organizers, PYA protesters made their way to front-row spots at strategic points along the parade route. Meanwhile, King, Stearns, and other

PYA leaders went over to Wheeler Avenue Baptist Church, which had been recently been opened by Reverend Bill Lawson, a TSU Bible professor. Lawson "was on the phone," King remembered, with a direct line to the "conversation with the head of the Loews Theaters." The church had two phone lines and the other rang incessantly with reports "from downtown about their readiness to step in" and block the parade. The protesters kept calling, "asking whether or not it was a 'go.'"

On the way down to the parade route, the protesters had all kept an eye out for the nearest pay phone because, as King remembered, they "had to get to a public phone to call back to where we were at headquarters to find out what they were to do." Once they had word, they passed it along to the other forty or so protesters, usually by running the length of the parade route and whispering in their ears. It was a cumbersome operation, requiring painstaking planning. Word had to get out, it had to get out to everyone, and it had to be right. A mistake could disgrace the city or disgrace the movement. "We had a failsafe point up to perhaps—say—twenty or thirty minutes," King said, "and they would not be in contact with us beyond that point." Talks had to end by then. If not, "even though they may have settled then, we could not stop the protest."

The parade was set to begin at 11 a.m. By 10:30, the marching bands were gathering. Cooper and his family, local politicians, and NASA dignitaries and their wives all headed to their open-topped cars. The staging area was set. Children holding American flags were sitting down with their parents on curbs lining the route. People were lining the upper-story windows waiting to shower down paper. Negotiations between Jones, Taylor, and the people at Loews continued.

Finally, at 10:40, "maybe twenty minutes before that failsafe point," Mease picked up the phone and called over to Wheeler Avenue Baptist Church. The PYA had won. Mease told Stearns, "Jones had told Taylor that theaters and restaurants would be desegregated within 30 days. Stearns and King canceled the demonstration."

The negotiators put another secret plan in place. As had happened with the lunch counters, "these facilities would be visited by very small numbers of well-dressed, quiet and orderly blacks. Jones and Dundas again made sure that there would be no newspaper, radio or TV coverage." At a meeting at the Y, Mease passed out movie tickets. "He and Stearns went to see Steve McQueen in *The Great Escape* at the Loews

Theater. High school principal Arthur Gaines and his wife, Jeannie, dined at Pier 21. There was no trouble."[62]

Until the day he died in 2012, Otis King considered the threatened Gordon Cooper protest as one of the greatest triumphs of his time working for civil rights. His family even alluded to the event in his obituary. Although there are those who argue that the Gordon Cooper parade was the last hurrah for the PYA, Otis King did not remember it that way.[63] He saw that day as a pivot point. "We started out, I guess following what had happened in [the sit-ins in] North Carolina with a sort of a narrow focus. We took the easy targets." With the Cooper parade, the PYA took its game to a completely new level. NASA did not have the footprint in Houston that it had in Huntsville, so the opportunity to exploit it for the benefit of the African American community was not readily available, but it took some guts to threaten to destroy the welcome-home parade of an astronaut hero and publicly tarnish the city's reputation to bring about change. After Loews opened the movie theaters, King said, "our thinking shifted to jobs and other issues."

TSU students, along with students from the University of Houston, later organized two thousand African Americans for an April 1965 protest march against the slow pace of Houston's school desegregation. As with the Gordon Cooper march, the protesters turned Houston's claim of Space Age progress against the city, carrying signs that read "Space Age Houston, Stone Age Schools."[64] And they did not stop there. The movement, King said, "started with a quest for integration of facilities; that moved from there into employment and then moved from there into politics."[65] The movement changed Houston, and like so many other struggling civil rights efforts, it found strength by putting the Space Age promise of the future alongside the segregated reality of the present.

Houston was also home to the greatest living symbols of the American Space Age. The astronaut corps was an all-white, all-male bastion, making it a target for civil rights advocates and social critics. Almost as soon as the American space program notched its first successes, the black press and its audience began to call for African American representation in that elite company.[66] Sometimes the voices were subtle, as when the *Atlanta Daily World* congratulated John Glenn on becoming the first American to orbit the Earth. "May this nation under God," the paper said, "come home to the common reasonableness that all of her people share in this mighty feat that has startled all Christendom; that all her people under one flag must be heard and respected."[67] Sometimes they were harsher and more direct. A group of all-white astronauts, said the *Daily Defender*, was just "the same old story that's being repeated in the new frontier."[68] If the Space Age was truly a new era, the black press said, and if NASA was the embodiment of that era, and if the astronaut was the personification of NASA, then it was only right and fair that NASA find a way to enlist a black man in the astronaut program. Otherwise, as columnist Dan Burley said, "NAACP, CORE, United Sons and Daughters of Georgia, the Associated Nieces and Nephews of West Mississippi, the Amalgamated Reverends Inc., and the Thunderbird Brigade should be sending their assembled pickets over the ramparts of Cape Canaveral on the coast of Florida where Uncle Sam has cruelly and callously thumbed his nose at equality for those not white."[69] In 1962, the black press appeared to get its wish in the person of Edward J. Dwight.

First of Race in Space
Ed Dwight

As time passed, I fully understood that this was not an effort to get me into space, but was an effort to solely inspire Black kids to get educated.

ED DWIGHT

E dward J. Dwight was unquestionably the most prominent African American associated with the Kennedy-era space program. In 1963, his image graced the covers of countless magazines and the front pages of every black newspaper. He was to be the "First of [His] Race in Space"[1]—possibly even the first man on the Moon.[2] With perspective, however, his story is much less clearcut. In fact, Ed Dwight was never really part of NASA's astronaut program, and while it is clear that someone at the highest levels of the federal government wanted him to eventually be there, it is impossible to know who or why. The only thing that is clear is that at some point in the early 1960s, someone decided America needed a black astronaut.

When the public at large turned its focus to NASA and contemplated the new skills, new ways of thinking, and new people who constituted the agency, they did not think first about technicians, engineers, and scientists (or the bookkeepers, graphic artists, typists, public relations people, and others who helped make NASA run). Instead, the personification of NASA in the public's mind was the astronaut. The astronaut was the Space Age incarnate, the conqueror of the New Frontier, and the modern embodiment of the American pioneer spirit. When he flew into the heavens aboard a Saturn V rocket, and the words "United States," emblazoned across the side of its 364-foot exterior, streamed past the camera into our living rooms, he represented us all.

The astronaut corps, however, was untouched by civil rights–related executive orders and so, in reality, astronauts represented only one race and one gender. Early astronauts were military personnel assigned to special duty at NASA. Technically the agency did not employ them and had no apparent legal obligation to integrate this most exclusive club.[3] Nonetheless, in July 1961 the White House requested information on the possible selection of minority astronauts after NASA and the air force announced the fifty test pilots who would take part in future joint space missions. In a memo a few days after the announcement, Frederick Dutton, special assistant to the president, wrote to Adam Yarmolinsky, special assistant to the secretary of defense and the Pentagon's voice on the PCEEO. Dutton said it was important "that for symbolic purposes in crossing the frontiers of space, this country would have qualified members from minority backgrounds." Doing this, he said, "would find great response throughout the world." In his reply, Yarmolinsky confirmed the selection of the fifty test pilots and offered a brief explanation of U.S. Air Force (USAF) selection criteria. Candidates, he wrote, were either graduates of the USAF Experimental Flight Test School or nongraduates with "an engineering degree, combat experience and were currently qualified in Century series fighter aircraft."[4] In either case, he said candidates had to meet performance standards prior to selection. According to Yarmolinsky, USAF criteria did not exclude qualified minority applicants.[5]

Include Representatives of Minority Groups

Someone at the White House found Yarmolinsky's answer lacking, and Dutton advised him that minority recruitment by the air force of space mission candidates was desirable not only for public relations but also as a reflection of administration efforts "to include representatives of minority groups in significant undertakings." Dutton requested a follow-up report that was to include specific information on minority recruitment and the selection process. Furthermore, he asked Yarmolinsky to provide a target date for the selection of a minority candidate. November 1, 1961, was the deadline given for gathering specific information on minority candidates.[6]

In his September 13 follow-up report, Yarmolinsky wrote that the air force and the office of the secretary of defense were acting to identify minority candidates, but that the possibility of finding one by the target date was minimal.[7] In fact, a memo sent to the directors of manned spaceflight a few weeks earlier said, "No Negro applicants are in the 32 who have survived screening to the point at which we stand today." The military had sixty-three men in the initial pool of applicants. The only African American in that group was a member of the army with past disciplinary problems "severe enough to result in reduction in rank" who had not been passed through.[8]

While the special assistants exchanged memoranda, NASA administrator James Webb received a letter from Edward R. Murrow, director of the U.S. Information Agency, urging NASA to "put the first non-white man in space." Selecting a black astronaut, he said, would pay valuable dividends for America in the "non-white world." Webb answered that the current complement of astronauts was sufficient for agency plans at that time but that he would keep Murrow's suggestion in mind for future projects. Webb received a similar suggestion from the Reverend Uriah J. Fields of Montgomery, Alabama, in March 1962. Fields reportedly told Webb that black Americans took pride in the astronauts' achievements but wanted to see a black man in the next astronaut selection. Webb and NASA maintained that they based their selection of the original seven Mercury astronauts and future spacemen on technical qualifications and program requirements. The administrator insisted that the agency considered qualified applicants without regard to race, color, or creed.[9]

"Bobby Kennedy Wants a Colored in Space"

In early 1962, either President Kennedy ordered, or his brother, Attorney General Robert Kennedy, strongly suggested, to General Curtis LeMay, chief of the air force, that the USAF Aerospace Research Pilot School (ARPS) at Edwards Air Force Base, California, be integrated.[10] This communication is what placed Ed Dwight on a career track that could have sent him into outer space. The legendary Chuck Yeager, who in 1947 became the first man to break the sound barrier, ran the school at that time. Yeager had his own version of how Dwight came to enter ARPS, and his story matters because of the way Dwight's saga ended.

According to Yeager, Dwight's class consisted of test pilots "whose abilities and academic background were demonstrably outstanding." Yeager said he and his staff had winnowed down the list of cadets to a final group of eleven and that after they published the list, he received a phone call from the chief of staff's office "asking whether any of the first 11 were black pilots." Yeager said no. Dwight had been on the longer list of cadet candidates, but he had missed the final cut, coming in twenty-sixth out of twenty-sixth. Soon after, Yeager got a call from General LeMay. The general, referring to the attorney general, told him that "Bobby Kennedy wants a colored in space. Get one into your course."[11] Backed into a corner, Yeager imposed on LeMay to increase the amount of money the air force spent on the astronaut class so that he could expand from eleven students to fifteen and include Dwight as one of them. He said LeMay agreed; they increased the size of the class, and Dwight was in.[12] Dwight arrived at Edwards in August, an event celebrated by the black press. The *Cleveland Call and Post* trumpeted that he might become the "First of [His] Race in Space."[13]

Dwight was born and reared in Kansas City, Kansas, in a middle-class Catholic family. In 1945, Dwight's sister integrated Kansas City's Ward Catholic High School.[14] This was nine years before school integration in Topeka, sixty miles away, would result in the *Brown v. Board of Education* case. At Ward Catholic High, in a compromise with angry white parents, the bishop agreed to allow Dwight's sister to attend school, but Dwight would have to wait a year. That delay would allow the school to build a new, black-only shower facility and, Dwight wrote later, to "give them time to lecture and orient me on staying away from the white girls."[15] After high school, Dwight was inspired by a story in the black *Kansas City Call* newspaper about a pilot named Dayton W. Ragland who, like Dwight, was short and from Kansas City. Dwight tried to enlist in the U.S. Air Force. They rejected him because of his height, because he had a considerable speech impediment (he stuttered), and because he was black. Dwight said that he wrote to the Pentagon and got a letter back giving him a date and time when a Pilot Selection Team would be at the junior college he attended. He passed his tests and the air force sent him to Lowry Air Force Base in Denver.[16] He received his commission in 1955, the same year that Arizona State University awarded him a bachelor's degree in aeronautical engineering. At the time of his pre-astronaut training, Dwight was a twenty-seven-year-old B-57 pilot with 2,200 hours of flight time, including over 1,700 hours in jet aircraft.[17]

During Dwight's eight-month phase 1 training at Edwards AFB, the would-be astronaut said he faced racial prejudice and also resentment because of the perception that President Kennedy had intervened on his behalf and placed him in the program ahead of better qualified white applicants. Yeager told a different story. He called Dwight "an average pilot with an average academic background." Under the circumstances, competing in a class with the very best of the very best, Yeager said there was no way Dwight could keep up. "In those days, there were still comparatively few black pilots in the Air Force," Yeager wrote in his autobiography, "but Dwight sure as hell didn't represent the top of the talent pool. I had flown with outstanding pilots like Emmett Hatch and Eddie Lavelle; but unfortunately, guys of their quality didn't apply for the course. Dwight did."[18]

"U.S. Trains First Negro Astronaut"

Dwight graduated from phase 1 in April 1963 and qualified, with fourteen others, all of them white, for phase 2—the Aerospace Research Pilots Course (ARPC). He said of his accomplishment, "I hope this will give others of my race more incentive to apply for future programs." The *Advertiser*, a white paper in Montgomery, Alabama, praised Dwight's success and urged white Southerners to wish him luck.[19] Whether or not whites complied, the response in the black press was cathartic. A picture of Dwight in his military uniform appeared on the cover of *Jet* magazine. The cover of *Sepia* magazine showed him in a flight suit next to a jet with the legend, "U.S. Trains First Negro Astronaut." His hometown black paper called him the same thing. Finally, the black press was saying, African Americans are part of the club; America is sending a black man into space and we truly are all equal in the Space Age. The *Daily Defender* went so far as to suggest that Dwight might be the first man on the Moon. So did *Sepia*. The *Defender* toned down its coverage a few days later to say, just as excitedly, "No Racial Barriers for Man in Space—Astronaut."[20]

Capitalizing on this explosion of national celebrity, according to Dwight, Whitney Young, head of the National Urban League, set up "thousands of speeches I made around the country to every Black group, school system, & white professional organizations," which showered him with awards.[21] In his speeches, the absence of barriers was a common theme. His presence in the ARPC, he told audiences, showed that in America any man, regardless of his race, could reach for the stars if he had the ability and worked hard. Hard work was the principal theme when he spoke to African American schoolchildren, which he did quite a lot. In fact, according to Dwight, Whitney Young thought the whole idea of having a black astronaut "was *not* an effort to get me into space, but was an effort to solely inspire Black kids to get educated."[22]

To help in that process, an African American college professor, Dr. Charles Lang, produced a filmstrip that NASA used extensively as part of its Spacemobile Program, a traveling educational vehicle that went around to inner-city schools. The first half of the filmstrip infused kids with the excitement of a space launch, the roaring rocket engines and clipped, fast-paced crosstalk between John Glenn and Launch Control. What followed was a pitch directed to African American schoolchildren: study hard and you too can find a place in the excitement of space exploration. "The future," the narrator said, "will see many boys and girls like you growing to design better space ships, creating better cities and making many things better than they are today." Two African American elementary schoolchildren, Keith and Angel (played by Dr. Lang's son and oldest daughter), are then shown visiting NASA contractors, where they learn that African Americans work in the space program. The capstone of their trip is a visit to meet an African American hero, the man who the filmstrip insinuates will be the nation's first black astronaut. "Captain Edward Dwight," the narrator says, "was selected to be a part of the astronaut training program conducted by the United States Air Force. . . . The children meet Captain Dwight and his boy and girl on the wind-swept airbase where he received a part of his special training." There are pictures of Dwight walking around the various aircraft at Edwards Air Force Base, and at the end, "Keith is so impressed by his visit to the airbase that he asks Captain Dwight, How does one become an astronaut?" Dwight's voice is heard, telling him, "Well Keith, it's like I told my own kids: one must study hard in order to get anywhere in life." When Dwight finishes talking, the narrator says, "Angel listened with interest and then wanted to know whether or not girls too would have a chance to take part in the space program when they grew up." Dwight tells her,

"Our country is going to need both boys and girls with knowledge, imagination and courage to make it even greater than it is today." "This answer," the narrator says, "reassured her."[23] The filmstrip had a profound impact on children who watched it. Roscoe Monroe, who ran the Spacemobile Program, said, "I know what I saw in the reaction of Black youngsters who looked at Captain Dwight and how they reacted to his message—it did a lot of good. He was the closest person to an astronaut our group could identify with."[24]

While Dwight's message to African American children was about working hard, to African American adults his ascendance became a symbol of defiance against discrimination. In the fall and winter of 1962, Greenwood, Mississippi, was the site of extensive violence by whites against blacks during a voter education project run by the NAACP, the Congress of Racial Equality (CORE), the Student Nonviolent Coordinating Committee, and the Southern Christian Leadership Conference. Violent acts included, but were not restricted to, arson, attempted murder, and the unleashing of police dogs into crowds of black demonstrators. Just as Montgomery and Selma soon would, Greenwood became synonymous with white resistance to civil rights.[25] In the midst of the Greenwood violence, the *Baltimore Afro-American* ran an editorial cartoon. It showed two white men, one standing on a prone black man, above the words "Greenwood, Miss." The white men are staring up in the sky, confused or frightened, at a rocket ship, carrying a smiling Ed Dwight, flying overhead; Dwight's name and rank are emblazoned in huge letters on the side. The caption on the cartoon reads, "How in the World Can We Still Keep 'Em Down?"[26] The black press and its audience had their hero: the man who offered their ticket to the Space Age and all that term conveyed as far as solving the problems of today tomorrow. To paraphrase those editorials that ran in the black press after John Glenn's launch, no longer was Uncle Sam cruelly and callously thumbing his nose at equality for those not white. Ed Dwight showed America that all of her people, under one flag, were being heard and respected.

The Dwight Case

According to the deputy commander of the Aerospace Research Pilot School, Captain Dwight was the "logical guy" if it was important for the country to have an African American in space. Apparently, NASA felt it was not important. Dwight graduated eighth in his class of sixteen from phase 2 of the ARPS and was one of twenty-six astronaut applicants recommended by the air force to NASA. In 1965, NASA selected 14 of an initial 136 applicants for further training, two from Dwight's phase 2 class—but not Dwight.

The *Houston Informer, Baltimore Afro-American*, and other black papers ran extensive stories on Dwight's departure from astronaut training. The *Informer*'s banner headline, "NASA, Pentagon Deny Charge" was larger than that later used to announce the Apollo 11 lunar landing.[27] At the time, Dwight absolved NASA of any racial bias in its decision, saying, "That was entirely up to NASA. They had their needs and criteria, and I respect their decision."[28]

Dwight's defense of NASA is subject to interpretation. First, NASA might well have treated him as the equal of the white pilots and passed over him on merit. After all, there were far more pilots than astronaut job openings and many whites, just as qualified as Dwight, were passed over by NASA. A second reason for Dwight's favorable disposition toward NASA was his continued interest in becoming an astronaut. Whatever Dwight's reasons in the moment, though, his charitable attitude soon changed. The air force sent Dwight to Wright-Patterson Air Force Base in Ohio and gave him a desk job. In Ohio, Dwight said he suffered discrimination trying to find off-base housing and that racist neighbors threatened his family by screaming at them and throwing rocks through their windows. Angered by his life circumstances and his perception of how the air force and NASA had treated him, Dwight wrote a fifteen-page report and circulated it well outside the normal chain of command, sending it to Members of Congress and congressional staff who had been sympathetic to him, as well as to the White House. The report made its way onto the desk of the secretary of defense and into the black press.[29]

Ebony magazine associate editor Charles Sanders collaborated with a reporter from the Washington bureau of the *Philadelphia Inquirer* to write an investigative story about Dwight's case and his situation. The piece reads in parts like a Cold War spy novel: "'I've been warned about talking to reporters,' the Pentagon man said over the telephone, 'but there's smoke here and there may be fire. Let's go with it.' There developed in a series of late-night telephone calls the official's

NEGROES WHO HELP CONQUER SPACE

Over 1,000 Negroes are in satellite, missile field

FROM the blockhouse came the final countdown: ". . . five-four-three-two-one-zero." Seconds later, flames spewed under the Jupiter-C, then came the lift-off that sent Explorer I roaring into space above Cape Canaveral, Florida. In a hangar that served as a viewing point a mile from the launching pad, 48 whites and two Negroes, members of a transportation crew which had played an important role in the historic drama, celebrated wildly. "Boy," they shouted to each other, "it's up there. We did it! We did it!"

And so they had. Hauling fuels, tools and other materials and instruments for the firing, they had performed their jobs at their level in their own special way. And on other levels, in a very special way, top-flight Negro scientists, satellite and missile experts are performing brilliantly all over the nation in answer to the world's most exciting challenge—the conquest of space. This challenge of space and military preparedness has prompted a quest for scientific and technical know-how that transcends the traditional barrier of race. At least 1,000 Negroes have already responded.

EBONY's survey of 100 leading industries engaged in this work, reveals that Negroes are not Johnny-come-lately's in the field of satellite and missile research. A number of these industries refused, usually for security reasons, to give information concerning their Negro personnel. Like the Armed Forces, private industry included promising young Negroes in their training programs a decade ago. Today, those men are making outstanding contributions to research and development, and are responsible for much of the basic research in chemistry, cryogenics, radiology and the earth and life sciences.

In mathematics and atomic physics, for example, J. Ernest Wilkins Jr. is probably the country's foremost researcher, regardless of race. No less regarded is the work of Dr. Warren Henry at the Naval Lab in Washington, D. C. His research and knowledge of materials at extremely low temperatures is probably unsurpassed in the U. S.

Jupiter-C rocket which lifted first U. S. satellite, took off from Cape Canaveral, Florida, last January. Work of many Negroes helped make launching possible

Continued on Next Page   19

From the beginning of the Space Age, the black press strove to demonstrate that African Americans belonged in this cutting-edge field. This was the first page of an article in Ebony *in 1958. Courtesy Johnson Publishing Company, LLC. All rights reserved.*

Julius Montgomery presents the 2013 Florida Institute of Technology Julius Montgomery Pioneer Award to instructor and counselor Barbara Moore. Photo by Florida Institute of Technology Office of Alumni Affairs.

Julius Montgomery (front row, center) with the RCA softball team at Cape Canaveral in 1968. Ten years earlier, on Montgomery's first day at the Cape, he said, "Nobody would shake my hand." Photo by USAF AMR Cape Canaveral. Courtesy Jack Tolman and Vic Craft.

The Clifton Schoolhouse (top) was built by Butler Campbell in 1892 to teach the children in the free black community of Allenhurst. The building and the rest of the town was destroyed in 1963 to make way for the expansion of Cape Canaveral and the construction of NASA's Vehicle Assembly Building (bottom). Photo on top from North Brevard Heritage Foundation; photo on bottom, NASA.

Frank Crossley, the first African American to receive a PhD in metallurgical engineering, was an early beneficiary of affirmative action. He went on to invent a new class of titanium alloys. Family photograph courtesy of Frank Crossley.

No Comment

As early as 1961 the black press recognized the growing gap between the technological progress and symbolism of the space race and the lack of progress in race relations. This cartoon appeared in newspapers on May 20, two weeks after Alan B. Shepard Jr.'s spaceflight on May 5 and days before President Kennedy declared a Moon landing to be a national priority. Cartoon by T. Stockett, courtesy of the Afro-American Newspapers Archives and Research Center.

Ed Dwight qualified with fourteen others, all of them white, for phase 2 of the Aerospace Research Pilot School (ARPS), a stepping stone to becoming an astronaut. Dwight never became America's first African American astronaut, but in 1962 the black press treated him as if he would. Photo courtesy of Ed Dwight.

In 1962 the black press used Ed Dwight's ascendance to pre-astronaut status as a symbol of African American defiance of discrimination. Cartoon by T. Stockett, courtesy of the Afro-American Newspapers Archives and Research Center.

account of The Dwight Case from the day it began in July 1961."[30] The story hints at pervasive racism and cover-ups at all levels. Many in the military who had been responsible for Dwight being passed over were infuriated by his report—that he wrote it in the first place, that he did not follow the chain of command, and that the report made its way into the public. Chuck Yeager was so angry he told the air force's chief lawyer that he wanted to charge Dwight "with insubordination. If he brought charges against me and couldn't make them stick, I want that guy court-martialed."[31] That did not happen and in short order the Dwight controversy was off the front page.

Dwight's problems occurred at the same time as Gemini 4 and Major Edward White's spacewalk, the first for an American. The Gemini mission's success smothered coverage of the Dwight case, though the Soviet newspaper *Tass* included a reference to Dwight in its coverage of Gemini 4. When Gordon Cooper, on a goodwill trip to Kenya after his Gemini 5 flight, told reporters that NASA had never found an African American qualified to be an astronaut, Dwight shot off an angry letter to the president. Dwight's connection to the Johnson White House, aide George Reedy, had just stepped down as LBJ's press secretary, and it is impossible to say whether the president ever saw the letter. The White House forwarded the letter to NASA and the air force, which wrote back to Dwight. Soon after, Dwight wrote President Johnson of his plans to leave active duty.[32] After three years, the story of "astronaut" Ed Dwight was over.

Overall, the evidence suggests neither that NASA had orders to integrate the astronaut corps—if Kennedy gave an order, he gave it to the air force—nor that the agency had an official, institutional prejudice against black astronauts. In any case, NASA and the Johnson administration clearly fumbled a spectacular public relations opportunity for both the space program and federal civil rights policies. Captain Dwight's popularity in the black community was real. Dwight was a space hero to blacks equal to the white John Glenn.

For those who debate "the Dwight case," as many call it, it is difficult to get to the bottom of the story. Dwight sympathizers tend to take his story at face value and see a man beset by a culture bent on doing anything it could to keep him down. It is easy to overlook the resentment that instructors and others at ARPS must have felt over a man who got into the program through political connections. Chuck

Yeager said that the selection of that first class of eleven "added tre-
mendously to the prestige of our new school."[33] It is also easy to ignore
the habit in Dwight's self-published autobiography of painting everyone who acted against him (and even some people who helped him) as
pathological, evil, racist, part of a conspiracy, and sometimes all of the
above. In his book profiling African Americans in aerospace, J. Alfred
Phelps' chapter on Dwight suggests that Yeager was able to cow and/or
bamboozle "a group of black civil rights lawyers and congressmen" who
came out to Edwards to look into what happened to Dwight. Phelps
acknowledges Yeager's assertion that the lawyers were "shocked to see
[Dwight's] poor grades," but he suggests that Yeager controlled the
training program with such an iron grip that he was probably able to
induce those grades by intimidating the faculty.[34]

Maybe Ed Dwight was wrong to think everyone was out to get him.
Perhaps there were just some people out to get him. After all, there
were people who did not want Ed Dwight to be an astronaut. Some of
them no doubt were racists. Others saw a hot head with sub-par grades
who got in scrapes he could not explain to their satisfaction and who
liked to go over the heads of his superiors when he was challenged.
Wherever the truth lies, too much is open to interpretation and at this
point too much time has passed to say who is right and who is wrong.

Curiously, no memo, letter, or interview with a principal illumi-
nates the chain of official communication leading to government intervention in the NASA astronaut selection process. No one can even say
if the president was the one pushing the idea of a black astronaut.
After all, LeMay told Yeager that Attorney General Robert Kennedy,
not the president, was the one who wanted "a colored in space." There
is, however, a story that appears on dozens of websites that mention
Ed Dwight and his story as "the first black astronaut." The wording on
Encyclopedia.com is typical: "The administration of President John F.
Kennedy, at the suggestion of Whitney Young of the National Urban
League, selected Dwight to become the first black astronaut trainee."[35]
The sole source for this version of history is Ed Dwight himself. Actually, the story on the web is a mangled remnant of Dwight's version of
the story, which is even more tantalizing and goes even further. What
he posits is an origin myth, a story of great men and martyrs who
cooked up the idea of a black astronaut in order to save the next generation of African American youth.

According to Dwight, in 1959, when Senator John F. Kennedy was running for president, he had a meeting with four titans of the civil rights movement: Martin Luther King Jr., Whitney Young, Roy Wilkins of the NAACP, and A. Philip Randolph, who, as head of the railroad porters union, had pressured President Roosevelt into making equal employment opportunity a priority. Dwight said that at one point their conversation came around to finding a way to get young African Americans involved in the coming Space Age by interesting them in pursuing careers in science, math, and engineering. "We needed to have parents, teachers, and the youth themselves involved," he said the leaders told Kennedy. "How would one get the Negro community on board? Whitney Young's solution was simple: 'Make a Negro astronaut!'"[36] This, Dwight said, is the genesis of the administration's actions that resulted in him being picked as a trainee.

While Dwight made clear that they did not specifically talk about him at this meeting of civil rights legends, he insisted it happened, and said, "It was *Whitney Young that confided that story to me.* I don't believe he had any reason to fabricate such a story out of 'whole cloth' & why he would tell it to me, anyway if it weren't true."[37] Whatever Whitney Young's motivation, Dwight's timeline shoots the whole story out of the sky. In 1959, no one would have thought to invite Whitney Young to the high-profile summit Dwight describes. King, Randolph, and Wilkins were eminent leaders of their community in 1959. Whitney Young was not. He was not the head of the National Urban League then. He was the dean of social work at a college in Georgia. Young would not gain national prominence for three more years, so there is no way he would have been asked by the leading lights of the civil rights movement to join them in a meeting with a possible future president of the United States.[38] Nevertheless, this story adds to the mythos of Ed Dwight and his role in the story of African Americans in the space program. For all the substantial accomplishments by African American technicians, mathematicians, and scientists to gain equality or at least a leg up for others in their communities, the top guns of the space program in the public mind were the astronauts. The first African American astronaut was the one history would remember. Yet, even as a non-astronaut, Ed Dwight left a story that outshone those of other African Americans in the space program, including those like the next one we will meet, who made significant contributions.

The View from Space
George Carruthers

*I didn't recall any riots as such; of course I may not have been
reading the newspaper every day.*

GEORGE CARRUTHERS

I f an alien life form happens to land on our moon a billion years
from now, it will find ample evidence of the six human visits
made there between 1969 and 1972. There are footprints, of
course, and lots of American flags, and over on the Moon's near side,
in the Descartes Highlands, near a photograph of astronaut Charles
Duke's family and a U.S. Air Force commemorative medallion, it will
also find the most significant scientific contribution made to the space
program by an African American. Coated in pure gold, it stands on
three legs and stares out blankly at the infinite void: the first telescope
ever set up on another planetary body, the Far Ultraviolet Camera/
Spectrograph/Telescope, created by physicist George Carruthers.

There are stories throughout literature about the weird kid—
diffident, shy, ultra-focused—who goes on to leave something special
to the world that spurned him as a youth. George Carruthers' story is
very much like one of those. He began studying the stars as a little boy
when his grandmother gave him a Buck Rogers comic book. "They
would talk about how they went to Mars and Jupiter and things like
that,"[1] Carruthers said, which "students can relate to better than someone writing some equations on the blackboard."[2] Carruthers grew up in
an educated family. His father, George Sr., was an engineer at Wright-
Patterson Air Force Base in Dayton, Ohio, and an uncle had a PhD
in romance languages and taught at Howard University and in New
York. College and careers were important in the Carruthers household

when George Jr. was young, and he remembered his father as "a role model in terms of giving me advice on studying math and science."[3] Unfortunately for George Jr., though, just at the time he was developing an interest in what would become his life's work, his role model moved the family away from the Cincinnati suburbs and into a tiny farming town. Milford, Ohio, was "really out in the boondocks," Carruthers remembered, and after the move, any advanced educational opportunities vanished. "I don't recall interacting with anyone who was technically oriented while we lived there," he said.[4] Young George was a self-starter, though, and being out in the middle of nowhere did not deter him from the pursuit of his interest. About the time of the family's move, he was leafing through his father's encyclopedia and found the section on astronomy. "The combination of that plus the science fiction magazines really sparked my interest," he said.[5] "Although the Buck Rogers comic books were good, [the encyclopedia was] factual," and he spent as much time as he possibly could at Milford's tiny library, seeking out and soaking up what he called "the reality parts."[6]

It is fair to say that in Milford, Carruthers was very much an outcast. He was the only African American child in his rural school, but it was not that. "I really didn't experience that much in the way of overt discrimination" because of race, he said. What made him an outcast, sadly, was his love of science. "Both teachers and students" in Milford, he said, "thought my interest in astronomy was strange." His family thought so, too, especially when in seventh grade he found instructions on how to construct his own telescope. He sent away for parts and spent hours first building the device and then standing out in the dark looking at the stars. "My father was sort of pressuring me to go into engineering," he said, and took a dim view of this turn of events. His mother's family also thought—like his teachers and fellow students— that he was strange. George persisted, though.

In 1952, *Collier's* magazine ran an article by Wernher von Braun entitled "Man Will Conquer Space Soon." Carruthers read that article, and for him there was no turning back. "There was no factual information on spaceflight at all up to that point," he said; but now his nighttime lingering with the heavens seemed to have a purpose. His father first interested him in math and science, but he had a new hero now. "Wernher von Braun was certainly a role model to me," he said. "I even wrote him a letter asking for some information and he sent me an autographed photograph as part of his response."[7]

When Carruthers was twelve, his life changed radically both for the better and the worse when his father died unexpectedly and the family had to sell their land in Milford and move in with his mother's relatives in Chicago. Carruthers had always hated Milford, and while he no doubt missed his father, his new life in the big city came with a significant upside as Chicago allowed him to immerse himself in the world of astronomy. The Sixty-Third Street elevated train ran right near his grandmother's home on the South Side, and it was a direct line from there to the front door of the Adler Planetarium. No one at the Adler was going to consider it strange that a young man wanted to spend all his time staring at the stars. On the contrary. "They had telescope-making classes," Carruthers said. Once he started high school, he began competing in science fairs, and "the very first year that I was there I actually had a telescope, one that I had built at the Adler Planetarium."[8] Carruthers reveled in his time at the planetarium. When the astronomers there scoffed at his dreams about spaceflight—"spaceflight was nonsense," they told him—he was not discouraged, and in fact their incredulity spurred him on. "I wanted to sort of counter that by studying aerospace engineering and making space astronomy a reality."[9] As Carruthers neared adulthood, the patterns of his life were coming into focus.

The African Americans who integrated the space program demonstrated a variety of coping mechanisms when subjected to overt racism. Julius Montgomery used his wits, rather than his fists; Frank Crossley—always the Good Sailor—rolled his eyes and powered through; and Theodis Ray steamed, only occasionally boiling over. Across that continuum, George Carruthers' behavior would fall on a far end, the best description probably being denial. He honestly contended that he never saw discrimination; that it never happened, even when it was all around him. In Milford, he said, in a sort of circular way, "they would make racial comments and things like that, and pick fights and things like that, but there wasn't a whole lot of that that I recall, at least not specifically racial in nature."[10] When he moved to Chicago, he said, "I used to ride a bicycle in white neighborhoods" and it caused a problem. "White guys came out and started shouting out to me," he said, but "I just gave them the same type of words that they used and kept on. Since I was riding a bicycle they couldn't catch up with me."[11] In college, he said, white people did not discriminate against him. They simply "don't want to actually socialize with you. So if you have study clubs that all get together and exchange notes, you're left out. Not that they would prevent you from coming, but they just don't invite you."[12] As he continued on, those blinders remained in place.

"I Didn't Recall Any Riots as Such"

Carruthers' love of science and engineering eventually brought him to Washington, DC. In high school, he read a book called *The Viking Rocket Story*, by Milton Rosen, which seemed to fuse his life's two greatest desires. Scientists, the book said, were using the Viking rocket to conduct astronomy. "That," he said, "was one of the major things that influenced my career choice." Herbert Friedman at DC's Naval Research Laboratory (NRL) was testing Vikings and launching them into space. At the same time, Carruthers' hero, Wernher von Braun, was also launching rockets into space in Alabama. The NRL research, though, involved "pure science," something "different from what von Braun's group was doing. The fact that the Naval Research Lab was interested in x-rays and ultraviolet astronomy was really great to me because that was something I could do that was new and different."[13]

Also at that time, African Americans in Washington, DC, were enduring something old and very much tried and true—a rigid system of Jim Crow segregation. When Carruthers arrived in 1964, blacks could not eat in downtown restaurants. Police stopped them randomly if they walked in white neighborhoods at night. Clerks would not let them try on clothing in downtown stores, if the stores let them in the door at all. Many blacks never bothered to leave the area around Howard University, fearing what would happen if they did. As with his time in Milford and Chicago, Carruthers remembered those days differently. "I don't think that there was really more segregation in DC than anywhere else," he said. Though he lived in the city through the devastating riots that burned whole neighborhoods to the ground following the assassination of Martin Luther King Jr., he claimed not to remember that the riots happened. "I didn't recall any riots as such," he said, adding, "Of course I may not have been reading the newspaper every day." The truth is George Carruthers had his nose to the grindstone all those years. It is reasonable to suspect that he never even bothered to look up at what was going on around him, and the irony is that he never looked up because his focus was on enabling others to look up—to the sky and the area beyond it.

The Far Ultraviolet Camera was the culmination of more than twenty years of work by Carruthers and a handful of other scientists to make it possible to observe objects from space—in this case to see the Earth's upper atmosphere from the Moon. "One of the things that Fred Whipple had espoused in one of these early *Collier's* magazine articles was the advantages of doing astronomy from space," Carruthers said. "In the 1950s [that] was not something that was well known even among astronomers, the advantages of going out into space."[14] Those advantages are significant, because once there is an eye up there, humans can see so much more. Looking through the atmosphere is a lot like looking at the lights at the bottom of a swimming pool filled with water. The water distorts the images, making the lights appear to flicker and move. It is the same with the stars when viewed from Earth.[15] While the Hubble Space Telescope can now look at objects outside the atmosphere, when Apollo 16 took up Carruthers' machine, seeing them was much more difficult to do. That is why he labored with such fervor to make it easier.

Carruthers' devotion to his science to the exclusion of all else— his refusal to see anything around him that did not involve the job at hand—was actually a common trait among those in his profession at that time. For a rocket scientist in the earliest days of the space program to shut out everything, to be "socially detached," was not aberrant behavior.[16] "One thing that I never did much of was actually socializing in the sense of doing anything outside of the laboratory," Carruthers said. The idea that he would be detached to the point of not seeing rampant discrimination aimed at him is also not a stretch, especially considering his proclivity to play down the racial slights he faced throughout his early life.

This detachment was not just a trait of Carruthers'. It may be one further explanation for why engineers "uncomfortable in a universe of words and feelings," like Julius Montgomery, buried themselves in their printed circuit boards, or why Frank Crossley, up to his neck devising innovative means of grinding polystyrene sheets, might be more likely to hear the N-word and let it slide.[17] In fact, it may also go far in helping to explain why the NASA science and engineering staffs had such a difficult time caring to apply the equal employment rules that headquarters considered so important.

Despite George Carruthers' inability to see it, the start of his career coincided with some of the most momentous events in civil rights history. They are also the times when the space agency and its contractors were as engaged as they would ever be in the efforts to erase racial discrimination, both in their own ranks and in the communities that housed them. The battle by African Americans to gain equal access to jobs, housing, education, lunch counters, toilets, water fountains, and the voting booth had been gaining national attention since the 1940s and 1950s. In 1964, Congress would pass the Civil Rights Act, and the following year they would pass the Voting Rights Act. After that, the focus would shift from access issues to economic justice, and the tactics would shift from marches and sit-ins to urban riots. In a way, the year before Carruthers arrived in Washington, 1963, was the year that the "old" civil rights movement ended. When it did, it did not go out with a whimper. From January through the fall, the nation stood by in shock as "clashes between the increasingly militant Negroes and extremist whites created an atmosphere of crisis."[18] Once the nation was in the middle of it, the Kennedy administration knew it could not be an impartial bystander.

On June 11, 1963, President Kennedy addressed the nation, declaring, "Race has no place in American life or law." Federal courts had said so "in a series of forthright cases," and his administration had enforced the concept "in the conduct of its affairs, including the employment of Federal personnel [and] the use of Federal facilities." He was now calling on Congress to act, preferably "at this session."[19] Kennedy, of course, would not live to see Congress take the "necessary measures" he called for. During his brief time in office, he instead took the tools allowed him by executive action and wielded the power of federal agencies to do what was possible. The navy, where Carruthers worked, had done its desegregation work in Frank Crossley's time. The rest of the space program, both NASA itself and its contractors, had considerable work to do. In NASA's case, there was a delicate balance to maintain. NASA administrator James Webb was an experienced political

operator. He recognized the need for NASA to maintain cordial relations with Congress. He also understood that an executive agency like
NASA was subject to the will of the president. Furthermore, he knew
that independent action by either an agency or its administrator might
embarrass the president and lead to political exile for the offender.

NASA's challenge to racism in its southern communities was com-
prised of several components: NASA centers would obey and enforce
presidential executive orders, laws passed by Congress, and directives
issued by NASA headquarters, applying economic pressure at local
and state levels, as intended by the federal government. In so doing,
NASA would help to convince and sometimes force business and civic
leaders to pursue desegregation. More importantly, however, NASA
officials used the agency's enormous prestige and public image as political leverage—most significantly in Alabama, as we will learn later.
Any suggestion that this was deliberate is misdirected. The government was not going to spend all of those billions of dollars just to make
a point. The mission of NASA was space. Because it was in the South,
NASA had the ability to make a difference in society as a whole, but
only if doing so did not get in the way of its principal mission.

Exceptionalism

From the very beginning of the agency, the people of NASA saw
themselves as a breed apart from those in other federal agencies.[20]
What principally made them different was a culture that included
normalizing risk, accepting failure, and anticipating trouble.[21] Those
traits certainly applied when it came to testing rockets and launching
humans into outer space. The traits seemed to disappear completely,
however, when it came to the job of recruiting a type of personnel who
fell well outside the era's accepted cultural norm. When NASA was
attacking that problem, the normalization of risk went by the wayside
and, more often than not, "bureaucratic caution and political interference" became the order of the day.[22] As for what accounts for this
dichotomy, ascribe it to NASA's tendency to promote to its top management positions scientists and engineers. These were people char-
acterized by "the narrowly technical focus of their education."[23] More
often than not, they saw time spent on management issues as time taken
away from important technological problems.[24] NASA grew out of the

"intimate, free-wheeling working environment thought conducive to engineering innovation" that had dominated its immediate predecessor, the the National Advisory Committee for Aeronautics (NACA).[25] When the regulations embedded in the 1958 National Aeronautics and Space Act imposed bureaucratic controls over that environment, many engineers chafed at them. The tension between headquarters and the engineers at the field institutions persisted. As a result, no matter how many times James Webb asked his managers and center directors to work on something "which may seem not to contribute immediately and directly to our program schedule," his team felt they knew better what the real priority was.[26] They were there to perform feats of technical engineering. Social engineering was somebody else's job.

There was another conundrum. The technical staff at NASA were not just run-of-the-mill scientists and engineers. According to the people who ran NASA, the agency—working through and sometimes around the government's bureaucratic tangles—had managed to hire the nation's very best engineers and scientists and also harness and deploy the best people from the nation's aerospace industry. The upshot was that NASA was an agency that believed in the "exceptionalism of its employees."[27] Therein lay a problem with its hiring of African Americans.

George Carruthers believed that "most of the people who are doing science and engineering are less likely to do segregation than people out on the street."[28] At NRL, he said, "I couldn't see that there was any racial discrimination at all. There was discrimination in the sense that the older established engineers and technicians were negative toward the young PhDs," but that was the only discrimination he saw there. Carruthers was not able to discuss whether there was discrimination within NASA because NASA never gave him a chance to find out. "I did submit applications to various NASA centers," he said, but NASA never gave him a job. "I think I applied for summer jobs at NASA centers," he said, but "I don't remember being accepted by any of them."[29] He is, of course, wrong about people in science being "less likely to do segregation than people out on the street," especially in the southern towns that housed the principal NASA facilities. This is another explanation for NASA's troubles recruiting African Americans in the early 1960s. What happens when an agency based primarily in the South and staffed primarily by southerners must recruit into their exceptional organization people whom their culture has seen for three hundred years as inferior or even subhuman? An agency in that position would hold prejudices that exclude any black person, because, after all, African Americans cannot be the best. It might also tell itself that because it wanted the best, it needed to look in places where it knew the best people were, to the exclusion of other places.

This is in fact precisely what happened when NASA set out in the early 1960s to find African Americans to fill its engineering and technology ranks. According to James Jennings, who spent thirty-five years serving on NASA's Council on Equal Employment Opportunity, "I don't think NASA went and looked where the people were." Jennings pointed out that "schools like Morehouse, Tuskegee" and others were graduating "mathematicians, physicists every day." No one "was born a rocket scientist and nobody graduated a rocket scientist," he said. "They graduated in math, chemistry, physics or something like that," and it was necessary "to go and show those folks that there is an opportunity for you if you want to come and work at a place like NASA."[30] It would take some time, according to Morgan Watson, one of NASA's first black engineers, before anyone was able to take Carruthers' hero, Wernher von Braun, and "point him in the right direction and [say], 'Don't look at Auburn or don't look at the University of Alabama or Georgia Tech or any of those places for black engineers. Go to the black schools.'"[31]

Behind the Government-Wide Ratios

In the fall of 1963, shortly after President Kennedy's assassination, Kennedy Space Center, the Manned Spacecraft Center (Houston), and the Marshall Space Flight Center in Alabama submitted long reports laying out the steps they had taken to increase African American employment. At year's end, NASA issued a final, agencywide report based on those facilities. It found that the centers had developed some effective programs. It pointed out, however, that the public and the government were judging NASA "not merely by the sophistication of the programs but by employment statistics," and at the end of 1963 those did not paint a rosy picture. Government-wide, African Americans represented 13.1 percent of the classified federal service and 19 percent of the wage board employees. At NASA, by contrast, "the comparable figures were 1.9% and 4.9%." The report found that "almost every NASA installation is behind the government wide ratios in each area."[32]

The memo laid out plans for 1964 that included giving attention to the places at NASA where "turnover and training provide maximum opportunity"—which meant secretarial and wage board jobs. The report also talked about looking into the best way to promote co-op and summer employment programs and reviewing the "existing rosters of eligibles" at NASA's various installations and passing them around to the other installations. It talked of using training programs sponsored by the National Urban League wherever possible, and conducting surveys, which, it was hoped, might "assure equal opportunity for training and promotion of minority group workers." The report also suggested giving greater focus to encouraging contractors to hire African Americans. "We feel that the contractors in these communities can do the most effective job in those areas as full-time members of the community," it said.[33]

In the period after President Kennedy's death, NASA operated with a considerably greater amount of what the engineers called "pogo" when it came to civil rights. In finding a place for African Americans, the agency oscillated and vibrated almost as uncontrollably as the early versions of the Saturn V rocket. Throughout 1964 and 1965, the agency, its leadership, and the region it principally considered its base would travel like Gemini—like twins. NASA leaders and the South would rendezvous and dock—sometimes with the region changing the agency and, less frequently, with the agency touching the region. Throughout this period, NASA leaders would make grand gestures that challenged segregation's most powerful leaders, often at the risk of severe cost to the agency and its mission. But just as often, NASA would shrink away from opportunities to cope with segregation and even to carry out its responsibility to enforce policies against it.

The Civil Rights Act and NASA

On July 2, 1964, right about the time George Carruthers moved to Washington, there was a ceremony in the East Room at the White House. There, a bevy of white legislators and Martin Luther King Jr. surrounded Lyndon Johnson as he signed into law the capstone of al-

most a decade of pushing and cajoling for the benefit of racial integration and racial harmony, the Civil Rights Act of 1964. It codified
the racial equality Johnson had championed, and, it is reasonable to
assume, he hoped the legislation would further his plan to bring about
social and cultural changes through greater access to employment.

Johnson was always careful in public to link President Kennedy
with the passage of the Civil Rights Act, and while NASA leadership
did the same, Julius Montgomery said he knows who deserves full
credit. "Before John Kennedy was assassinated, it was no way in hell
you'd get anything passed," he said. Of Kennedy, he said, "There was no
way in the world that he could get a bill passed. But LBJ—there were
all those crude methods he had to get people to vote." Johnson, he said,
was the only one who "could get those Deep South Rednecks to pass
that bill."[34] This is also the assessment of historian Robert Caro, who
has called Johnson "the great legislative magician. He passed bills no
one else could pass." However, at the time of Kennedy's death, the Civil
Rights Bill was trapped in the House Rules Committee, whose segregationist chairman had "no plans" to bring it up. Before his first speech
to Congress, three nights after the assassination, Johnson's aides urged
him not to emphasize civil rights. "What the hell's the presidency for,
then?" the new president was said to have replied.[35] "Our first priority is civil rights," he told them. "'We've talked about civil rights for
100 years. Now it's time to write it into the books of law,' and he immediately takes Kennedy's two bills and gets them started to passage,"
Caro said.[36] Otis King's counterpart in his Houston protests, Quentin
Mease, was close to Johnson as a member of Houston's black cultural elite. He concurs in these assessments. Congress passed the Civil
Rights Act, Mease said, only due to "a stroke of nature." Kennedy's
assassination placed in power a southern politician who "made the
statement—and it is a matter of record—that he would do whatever
was necessary to end Jim Crow-ism not only in Texas but in the South.
And he did that," Mease said, not just by passing the Civil Rights Act,
but during his years running the PCEEO as well. "There were many
cases of government contracts being issued and they would have that
proviso in the contracts that there would be no discrimination in the
hiring." Because of Johnson, he said, African Americans being considered for government jobs were more likely to be judged "based on

their fitness for the position" rather than the color of their skin. To cap it all off, Montgomery said, "then when he had it, he said, 'I'm going to be the president of all the people.' And he says, 'We shall overcome.' I almost fell out the chair when he said, 'We shall overcome.'"[37]

Passage of the Civil Rights Act allowed James Webb to reaffirm NASA's institutional commitment to racial equality. The act gave NASA greater enforcement and compliance powers with regard to its contractors and other recipients of federal space money, including colleges, universities, and training programs. It also included procedures for compliance investigations, hearings, and reviews. Although the Civil Rights Act superseded all previous NASA-originated orders against discrimination, it affected neither the objects of those orders nor the provisions of Executive Orders 10925 and 1114. Simply put, the act became an umbrella for all federal agencies.[38]

NASA's Balancing Act

At this time, Congress had literally just written these laws, and their enforcement was happening for the very first time. Looking at NASA's pattern of rendezvous and docking with its southern units provides an opportunity to see, in microcosm, people trying to figure out how to live their lives (or, in this case, run their agencies) within these new laws. NASA in 1964 and 1965 traveled back and forth between responsibility and reality. New laws forced the agency to make new rules. It made the new rules with the best intentions, but abiding by them, especially in the South, could cause monumental pain. The world was changing. Now that it was, NASA, like America, was trying to adjust.

That adjustment required a balancing act by Webb. The government had cobbled NASA together from numerous existing entities spread all over the country and drawn from all over the bureaucracy. There were the three labs—Langley, Ames, and Lewis. There was the Dryden Flight Research Center at Edwards Air Force Base, Marshall Space Flight Center (MSFC), the Manned Spacecraft Center (MSC, eventually called the Johnson Space Center), the Kennedy Space Center (KSC). Then there was the Goddard Space Flight Center, the Jet Propulsion Lab, the Michaud assembly facility, Wallops Island, the White Sands Test Facility, and the space center in Mississippi. When it came to parsing the agency's apparent schizophrenia on race relations, the separate centers often behaved "like rival universities, each with its own traditions and interests." Each center also had its own favorite contractors, its own committees of outside advisors, and its own "long-range plans and priorities."[39] Each also had its own means of coping with racial integration.

Managing this balance, a feat accomplished with a considerable effort and a considerable amount of bravery, fell to people deep within the recesses of the organization. One example of that bravery is evident in a letter Webb sent in 1963 to Paul Bickle, director of the Flight Research Center in California. In it, Webb commends Bickle for taking action that was extraordinary in its time but is now more difficult to imagine. Bickle stood up for a real estate agent who sold a house to a black family. It happened in Lancaster, California. The realtor was Byron W. Troth and the buyer was Harold Washington, an employee at the Flight Research Center.[40] Washington was living in Sun Village, the black neighborhood closest to Edwards Air Force Base, but according to his friend Henry Hearns, the pastor of Livingstone Cathedral, "Everyone tried to live as close to the base as they could get." Mr. Washington wanted to move to College Terrace, but College Terrace "was an area a black person would have had a problem moving into," Reverend Hearns said. "The realtors would not take you there, or into Lancaster and N. Palmdale."[41]

It was like that all over California, apparently, as George Carruthers could attest. There are two incidents where his blinders came off and he realized the race problem that surrounded him. On one he declined to give details, merely chuckling and saying, "When we had some projects down in Mississippi, they definitely had segregation down there." On the other, he is specific. It happened in California while he was working a summer job with Aerojet General Corporation, one of those places that had told Frank Crossley that he was "too advanced for a Negro" to receive further promotion. In Carruthers' case, he had just come to California and arrived at the home where he expected to live. He said, "[I] went to the place where I was supposed to be and I waited and a car came by and then went around and then after half an hour the same car came around and went out. And then the third time the car came around and they said 'Well, I hate to tell you this but we don't take colored in this place.'"[42]

Byron Troth was not that kind of realtor, however. He "was a man of integrity with great respect for the valued identity of others" according to his friend Dixie Eliopulos, who owned an escrow company in Lancaster.[43] So he did take Mr. Washington to College Terrace. He also sold him a house, though not without repercussions. "The realtors at that time did not want to be an outcast to the people who were able to buy the houses—the whites," Reverend Hearns said, and Mr. Troth received a considerable amount of trouble from customers and from his fellow agents on the Antelope Valley Board of Realtors and the California Real Estate Association.[44] That is why NASA's Paul Bickle stood up for him. Webb was writing to let him know that he knew of his courage, calling the effort to find Mr. Washington a home "an excellent example of affirmative management action on your part in support of the President's program for equal employment opportunity." Webb sent a copy of the Bickle letter to all of the other NASA field installations to serve as a guide in handling similar situations.[45]

"Basically Fair Was Fair"

At NASA, a man named Alfred S. Hodgson was responsible for tracking small skirmishes in the war for equality, like the one addressed by Webb's letter to Bickle. During his tenure at the agency, Hodgson was also able to engage in a number of such skirmishes himself. A man who believed in fairness and who believed equally in government's power to enforce fairness, Hodgson labored for years to make sure that NASA lived up to the commitments expected of it as part of the Kennedy and Johnson administrations' civil rights agendas. Hodgson came to James Webb's attention when they both worked for President Truman. Hodgson was a career civil servant, and when President Eisenhower created NASA, records suggest Hodgson was its first civil service employee.[46] He became director of management analysis in 1958 and assistant to the director of business administration in 1960.[47] NASA handed its civil rights enforcement portfolio to Hodgson after the creation of the PCEEO in 1961. Following the committee's first meeting, Webb told the program directors and staff officers at NASA headquarters and all directors of field installations that as a PCEEO member, he "was to accept direct personal responsibility for vigorous leadership in this agency to accomplish the purposes set up in the President's Executive Order."[48] In announcing Hodgson's ap-

pointment, Webb told his directors, "I will expect your fullest coopera-
tion" with his activities to enforce equal employment. From that point
on, the agency began its multiple, stuttering steps to enforce the administration's equal employment agenda.

According to Alfred Hodgson's eldest son John, the position suited
Hodgson. "My father was a strong believer in civil rights," John Hodgson recalled. "He believed in it as a matter of fairness." As the early 1960s
progressed, "we talked a lot about [civil rights] at dinner. My father felt
very strongly that basically fair was fair and everybody ought to have
the same opportunity. It was as straightforward as that."[49] A spate of
memos flew back and forth between NASA headquarters and NASA
centers in the South in the early 1960s enumerating the most contentious battles the agency fought within its own walls and with its host
communities. Nearly all these memos contain the admonition, "follow
up with Al Hodgson."[50] The problem Hodgson faced was not just how
to further the goal of equal opportunity employment, but how to do it
within the culture of NASA. It was his job to make sure NASA's willingness to endure risk extended to hiring practices that might threaten
someone's feelings that the agency was compromising its exceptionalism. Finding this balance and then striking and maintaining it were
Hodgson's responsibility. He also got involved in integration fights directly. Once, while on a NASA site investigation in the Deep South, he
dove into a public swimming pool along with a black colleague in front
of a group of town officials. "Basically, they integrated the town's segregated swimming pool," John Hodgson said, "and they were sending a
message that, if the town wanted NASA there, this was the way things
were going to be."[51] Hodgson also worked behind the scenes to pressure Alabama governor George Wallace to integrate Huntsville's public
schools. These activities deserve attention in the calculation of NASA's
and the space program's overall role in civil rights efforts in the South.

Heavy (Often Heavy-Handed) Government Pressure

Scoring how well NASA performed overall is a difficult assignment,
subject to nuanced opinions about what constitutes enough action,
or even what constitutes "action" and what constitutes "enough." On
a purely statistical basis, the numbers do not look good. The Marshall Space Flight Center in Alabama had only 52 African Ameri-

can employees out of 7,335 in 1963. Six years later, at the time of the Moon landing, the center had only eight African Americans in management jobs.[52] The other centers did not do so well either. The day Apollo 11 landed on the Moon, the Manned Spacecraft Center in Houston had twenty-one African American workers; Kennedy Space Center had five.[53] Even with those numbers, however, commentators at the time tended to give NASA considerable credit for making a difference. William Ellis, who wrote the "Space Crescent" articles in the *Nation*, credited "the social impact of the aerospace industrial renaissance" with changing race relations in the area stretching "from Houston to the east coast of Florida, and from New Orleans to northern Alabama."[54] The same assessment came from Herbert R. Northrup, who developed groundbreaking theories on race in the workplace in his years at the Wharton School of Business in the 1960s.[55] In his book *The Negro in the Aerospace Industry*, Northrup said NASA and the major aerospace companies in the Southeast "changed employment practices of the region in a major manner." Ellis was not shy about crediting federal equal employment policy for the change. He quoted an unnamed "young engineer" who was transferred by Boeing from Seattle to MSFC: "Pressure by the government for more moderate race relations is there." Northrup agreed with that assessment. By 1963, he said, affirmative action plans had become prominent in the once highly segregated aerospace industry. That prominence, he said, stemmed from "heavy (often heavy-handed) government pressure which motivates employers and keeps the problem in the forefront and constantly pushes the industry to take further affirmative action." He talked about activities of the sort taken by NASA in Alabama and at Cape Canaveral and said they had had "great success." In fact, he found that twice as many African Americans were working in office and clerical jobs in the Southeast as in any other part of the nation.[56] Northrup emphasized that last fact with an exclamation point (the only one in his data-heavy book), and it was deserved! The research Northrup conducted took place at a time when African Americans in the South could reasonably expect to hold one of only about twelve kinds of jobs. Yet in the aerospace industry, thanks mostly to the PCEEO's pressure on NASA, he found that "representation among the officials and managers, professionals, and technicians for Negroes in the southeast is nearly identical with that nationally."[57]

Did NASA successfully advance racial equality in its southern host communities? Yes. Even in the most restricted of applications, NASA influenced its hosts through policy, perception, politics, and the federal purse. Success varied with local conditions, but in each community, NASA's presence weakened or destroyed some aspect of racial segregation. A common lament since 1969 has been, "If they can put a man on the Moon, why can't they . . ." NASA did land men on the Moon and in the process made life in the South less segregated. The agency's grand national impact on civil rights was considerably smaller than its legacy in outer space. Nonetheless, NASA and many of its African American employees were able to create important, lasting changes.

One of those, George Carruthers' Far Ultraviolet Camera/Spectrograph/Telescope, will probably remain where it is standing for many millions of years. It is therefore surprising to learn Carruthers' opinion of why NASA chose his device to go to the Moon in the first place. "They didn't actually have what NASA does now," he said, with scientists subjecting their projects to vetting and evaluation before the agency selects them. "Nowadays," he said, "only one out of ten proposals ever gets funded. But back in those days, very few people had those kinds of proposals." Despite NASA propaganda about it being a scientific agency, Carruthers said, the agency's principal mission was "trying to beat the Russians, of course." When it came time for missions, he said NASA often "had a lot of empty space. So they asked for something—'What can we do on the Moon that we can't do on the Earth?'"[58]

In 1969, Carruthers submitted a proposal to NASA to place the camera he had built at NRL on the Moon so that it could look back at the Earth in ultraviolet light. Coincidentally, another scientist named Thornton Page suggested doing the same thing. "In his proposal," Carruthers said, "he actually made the statement that he thought that the NRL camera was the one that was best suited for the job."[59] NASA agreed and asked the two to work together, with Carruthers overseeing the development and calibration of the instrument and Page in charge of planning the observations. In addition to looking at the Earth from above the atmosphere for the first time, the camera allowed humans to "see the full extent of the hydrogen atmosphere. All of these were revealed in pictorial form for the first time."[60] As for why NASA picked them, Carruthers said the fact that he and Page were "at the right place at the right time, I think, is really the important point."[61]

Carruthers was in the right place at the right time in many ways. If the sit-ins, the riots, the PCEEO, the Civil Rights Act—all that activity Carruthers said he never saw—if none of that had happened, Carruthers would probably not have been teaching astronauts how to operate his telescope on the Moon. Maybe he would have been a chemist at a foundry. Most likely, he would have been teaching math at a black college in the South. Though Carruthers missed all those events because he was too busy, many elements coalesced to make sure outstanding African American scientists could achieve their dreams. George Carruthers was exceptional and, with his camera and subsequent projects placed on the Space Shuttle, he worked with an exceptional agency. That exceptionalism, however, extended beyond the realm for which the public knew NASA best, and included a considerable amount of exceptional work in the area of race relations as well.

"Huntsville,
It Has Always Been Unique"

Delano Hyder and Richard Hall

We here in Huntsville just had a different attitude and feel about integration, segregation, whatever. It was a different atmosphere.

Delano Hyder

[Protesting] doesn't become a part of your feelings. It just doesn't become a part of you.

Richard Hall

Just as there are "many Souths,"[1] so, in the 1960s, there were at least two Alabamas. One's reputation was notorious—the "Cradle of the Confederacy, this very Heart of the Great Anglo-Saxon Southland," embodied by its governor, George C. Wallace, who spoke those words at the same inaugural address where he drew "the line in the dust and toss[ed] the gauntlet before the feet of tyranny," declaring "segregation now, segregation tomorrow, segregation forever."[2] The Alabama Wallace celebrated was a place of billy clubs and bombings, of fire hoses, snarling canine teeth, and government-sanctioned murders, plus selfless acts of ingenuity and bravery large and small by the very people who were the objects of that abuse. The other Alabama's reputation was as decent, honorable, and principled. That Alabama called itself Rocket City, USA. It was Huntsville, the former Watercress Capital of the World, transformed into the intellectual and engineering center of the United States' space program. The town, with its sixty-piece symphony orchestra, its "specialty shops, foreign restaurants and even art galleries" and "parks instead of pool rooms," had always seen itself as separate from the rest of the state that housed it.[3] A question remains, however: though Huntsville did not vote for Wallace, in the main, what did it think about Wallace's ideas—were they ideas Huntsvillians shared or ones they shunned? A convenient window into that question does exist. It can be found in the common experiences of Richard Hall and his forty-year NASA co-worker, Delano Hyder.

Hyder and Hall were both Huntsville natives. They were both at NASA from its earliest days and saw the changes it brought. As African Americans, they both knew each side of Huntsville. Because they were locals, though, they also had a Huntsvillian's perspective on their hometown's duality. The two Alabamas were "like night and day—two different places," Hall recalled.[4] If we scrape below the surface, though, we can see what the parochial perspective chooses to miss. Hall and Hyder's stories show that beneath the sunny surface, Huntsville had plenty of the other Alabama in it, too.

Smug northerners are happy to point out that the personifications of the two Alabamas, Governor Wallace and Marshall Space Flight Center (MSFC) director Wernher von Braun, were a racist and a former Nazi, respectively.[5] However, one cannot overstate the positive influence on the city's population of the former German rocket scientists and engineers who lived among them. Once settled in their new home, the team dove into civic and cultural affairs. Among their activities were the orchestra, new highway design, and the establishment of an observatory with a 21.5-inch mirror. According to pioneer space writer and frequent von Braun coauthor Frederick I. Ordway III, von Braun "was the guiding light who laid the cornerstone for modern-day culture minded Huntsville."[6] While there is plenty of room for argument about whether the space program changed the South, according to Richard Hall one thing is clear. "It changed in Huntsville," he said. "It definitely changed in Huntsville."[7]

"Just Like Talking about Alabama and New York"

Julius Montgomery was born in Birmingham, Alabama, which he called "Jim Crow headquarters"; he experienced total racial separation growing up there. Richard Hall said the contrast between his home and Montgomery's, though they are only a hundred miles apart, could not be starker. "Just like talking about Alabama and New York," he said. "We talked to white people and chopped cotton with them, picked cotton with them." Although—and here is a parallel with Frank Crossley's desire to always put the best face on things—he did add that "culture-wise, everything was totally separate." That shift at the end of his characterization points to the dichotomy at the heart of Huntsville's self-definition. MSFC deputy director Harry Gorman once called the racial climate in Huntsville "probably one of the most objective and progressive in the South, if not in the country."[8] It will be clear by the end of this chapter and clearer by the end of this book that that was far from true. Compared to the other Alabama, however, it is reasonable to see where he could get that impression. Delano Hyder offered further evidence that Huntsville, while not Birmingham, was most definitely still Alabama. "We talked to the white people that we came in contact with," he said. "We didn't really get to see them that much, but the ones that we came in contact, it was a good relationship."[9]

The situation comes into even tighter focus the farther we widen out. According to Horace "Pap" Rice, a contemporary of Hall and Hyder, "in the '50s paths didn't cross, things were rigidly segregated and everybody knew their place."[10] And even Hall, when pressed, had a story that shows what this meant. "We had a city park in Huntsville and the black community was on three sides of the city park. We could not walk through the park on the way to school. Therefore, we had to walk an extra mile just to get to school."[11] Hyder remembered that walking to school "we had to walk on the railroad tracks to get there— had to go through the trash pile."[12] But it was Dr. Sonnie Hereford, a man who helped organize the 1962 Huntsville sit-ins, who offered the most poignant recollection. "When I'd walk to school," he said, "we sometimes had to go in a roundabout way to get there, and the Caucasian kids always rode in the orange and yellow school bus, and whenever they passed by, when two or three or four of us black kids were walking, the only thing I remember about the school bus was that it blew dust in our face if it was going fast, and if it was going slow, eggs and rotten tomatoes came from the windows of the school bus and hit us in the face."[13]

Jim Crow also marred the schools to which they walked.[14] In Huntsville, Hall said, "There was only one school in the city from grade 1 to 12 for blacks." The school's resources were meagre; as Hall remembered, "I had never seen a book or magazine or newspaper. Never seen one!"[15] Things were a little better for Hyder. "This black lady had in her home a kindergarten which the neighborhood kids would go to. So I knew about books and that type of thing." Still, it is little surprise that the white citizens of Huntsville completed a median of 11.8 years of schooling in 1960 while the black completed a median of 7.3 years; and the median white family income in Huntsville was three times as high as that for blacks.[16]

Along with the major affronts, Huntsville also had plenty of small-bore annoyances for its black citizens. According to Sonnie Hereford, all one had to do, if he or she was black, was look around to realize something profoundly unfair was going on. "Take the stores on the square where the clerks would make you wait until all the whites had been served. You'd go into one, and maybe a clerk would be waiting on two or three Caucasian people, and you'd stand there and stand there, and then three or four more Caucasians would come in after you, and the clerk would finish up with every single one before turning to you."[17] He found life in Huntsville to be as restricted as any other town in Alabama. "There was no black policemen, no black mailmen, no black clerks in the stores. That bothered me."[18] As the space program geared up and Huntsville began its lurch into the future, Hereford said the racial disparities became more evident. "We saw the money going into Redstone Arsenal and NASA," he said. "But little of that money seemed to come our way."[19] He and many others in town thought the time had come to make sure that changed. Hyder and Hall, on the other hand, were there with von Braun and the Germans. They did not see a reason to kick up a fuss. "We here in Huntsville just had a different attitude and feel about integration, segregation, whatever," Hyder said. "It was a different atmosphere." Because of that atmosphere, Richard Hall said that protesting "doesn't become a part of your feelings. It just doesn't become a part of you."[20]

"At Last, It Has Happened!"

The issue of segregation clearly did become part of the city in 1962. In Huntsville then and in Huntsville fifty years later, two distinctly different impressions exist about the impact and the severity of that protest—a citywide series of sit-ins by local students. Aggressively ignored at the time by the local paper (the black press called it a "news blackout"), the sit-ins barely registered with whites.[21] The *New York Times* described Huntsville's experience with lunch counter sit-ins as "quiet and orderly."[22] NASA spokesman Bart Slattery Jr. told a reporter from the *Nation*, "We've never really had a problem [with race relations] here in Huntsville." As for why that was, Huntsville Chamber of Commerce manager Jimmy Walker put it succinctly, telling the same reporter, "We've got a high class of niggers here for the most part"—a quote that offers a stronger whiff of reality than Walker probably realized.[23] Memoirs by and about the community's black residents and coverage in the black press demonstrate that while official reaction to the sit-ins was nowhere near as ham-handed and violent as in Birmingham, Montgomery, and Selma, Huntsville was no less resistant to change.

The first Tuesday night in January 1962, it snowed heavily in Huntsville. Wednesday dawned frigid and the Madison County superintendent of schools declared a snow day.[24] While little kids grabbed their mittens and sleds, a group of young people from Alabama A&M College and Council High School had another plan. They took advantage of the day off to make a trip down to the Heart of Huntsville shopping center, where, around lunchtime, they walked into the Woolworths and the Sears. They "occupied the lunch counter stools for 15 or 20 minutes"; the Huntsville sit-ins had begun.[25] "'Sit-Ins' Finally Hit Huntsville: Twenty-Three Jailed in 'Missile City'" was how the *Pittsburgh Courier* headlined its January 20, 1962, story about protests. "At last it has happened!" the paper said. "This famous 'missile city'—home of the mighty Saturn, the Redstone rockets, Dr. Wernher von Braun, and the great Redstone Arsenal—has been hit by the sit-ins, and 23 Negro students of Alabama A&M College and local high schools were arrested last week after the first seven days of demonstrations."[26]

Over the next seven days, according to stories in the black press, "groups from 50 to 100 students invaded the business district of this fastest-growing city in America and entered the stores, seeking equal service."[27] Contradicting Huntsville's image as a place where "the Gospel of Wealth had more disciples . . . than did the Gospel of White Supremacy," things got ugly rather quickly.[28] A few days into the protest, African American students William Pearson, Leon Felder, Bertha Burl, and Mary Joyner, as well as a white NASA employee, technical artist Marshall Keith, walked into Woolworths. Keith, a Huntsville native, ordered a plate of eggs, and when the waitress served him, he pushed it across the table to the black students. According to a story distributed nationally in the black press, "The Woolworth's manager—distraught, attacked the waitress; he asked, 'Why'd you serve him with all of them sitting around him?' 'They weren't there when I brought his order,' answered the waitress. 'Take a good look at that stupid face. If you serve him again you're fired!' continued the manager." The manager turned to Leon Felder, who was now eating the eggs Keith had ordered. "'Want more salt and pepper?' He tilted the shakers over the plate, smothering the food with condiments." When Felder continued to eat, the Woolworths manager picked up the ketchup bottle and started pouring ketchup all over the table in front of the girls. When they quietly passed out tissues and started wiping up, the article said the manager, "Drawing once more from his store of outraged Caucasianisms," grabbed the plate of eggs and smashed it on the floor.[29] If the episode had ended there, one might be inclined to agree with an assertion made by MSFC personnel officer Art Sanderson that people in Huntsville "were quite a bit above the accusations about civil rights" that were attached to them. Terror, violence, and intimidation, Sanderson said, "may have been true somewhere south of here. It was not true here."[30] The aftermath of the encounter at the Huntsville Woolworths suggests otherwise.

"Khrushchev Can Eat Here but I Can't"

A few days after the incident, Marshall Keith, the white NASA employee, was sitting at home when there was a pounding on his front door. He opened it to find two masked men with guns. They ordered him out of his house. The men blindfolded Keith, marched him to their car, and threw him in the backseat. They drove Keith to a remote section of the city, where they had him get out of the car and strip naked. Then they pulled out a can of pepper spray and doused Keith from head to toe. The kidnappers hit him in the head with a blackjack, got in their car, and drove away. Keith managed to walk to the nearby home of an African American family and call the police, who took him to the local hospital, where doctors treated him for skin burns.[31] According to Hank Thomas, a CORE organizer who went down to help run the Huntsville sit-ins, "We could not contact him after he was released from the hospital. I was told his grandma had been threatened, that he had resigned his job. He left town for New York City. That was the last we heard of him."[32]

Hyder, Hall, and the rest of Huntsville's black community knew all about Keith's story and other outrages. A man named Charlie Ray drove through the black neighborhoods in a van with a loudspeaker on top, keeping people abreast of the news, so people were certainly aware of the danger they faced. However, the organizers of the sit-ins, Dr. Hereford, dentist John Cashin, and the Reverend Ezequiel Bell (whom Hereford called "our version of Dr. Martin Luther King Jr." for his ability to fire up a crowd) pressed on, attacking the problem on several fronts. While Hereford marshaled pickets to walk in front of downtown restaurants with signs saying "Khrushchev can eat here but I can't" and "Rocket City, USA: Let Freedom Begin Here," Cashin and Bell "started talking to local whites in meetings in private homes to drink coffee and discuss race relations."[33] When Hank Thomas from CORE showed up, activities jumped into a higher gear, Hereford said. Thomas was a student at Howard University and, according to Hereford, had been a Freedom Rider on the bus that was bombed outside Anniston. Thomas's experience in Huntsville was another blow to the city's reputation as a place different from the rest of Alabama. While the Huntsville protests were in full swing, Eleanor Roosevelt invited Thomas to Washington to testify at a hearing she convened of the Committee of Inquiry into the Administration of Justice in the Freedom Struggle.[34] There he testified that the Huntsville police followed him "all the while. Sometimes [they] kept a 24-hour tag on me." He also related an incident eerily similar to that of Marshall Keith. On January 15, he said he noticed a strange odor in his car. "Someone had torn the battery leads, but we got the car started," he said. "After about a block I got a stinging sensation in the seat of my pants. It was very bad. A little later I was in so much pain, I couldn't concentrate on anything. We found a doctor and he smelled my coat and said, 'That's just like mustard gas.' I was put in hospital, under sedation."[35]

As the sit-ins and arrests continued, and NASA launched John Glenn into orbit, members of Hereford, Cashin, and Bell's newly formed Huntsville Community Service Committee (CSC) sat down for two meetings with city mayor R. B. Searcy. At the first, the mayor told them "no racial problem existed" as far as he could tell.[36] He told Hereford that because people of all races in town "stand in the same line at

the bank, they go to work, they come home," he had no idea why protests should continue or any changes should be made.[37] At their second meeting, the CSC members disabused the mayor of his belief that
Huntsville's "Negroes were satisfied," and reiterated a request made
the previous year by a group of white ministers that he appoint a biracial committee to deal with segregation.[38] "Our thinking was that
cities that have established biracial committees during times of trouble
usually achieved desegregation faster and more smoothly than others,"
Hereford wrote. The mayor by now acknowledged that the city did indeed have a problem, but told Hereford he would not be able to get
white people to serve on such a committee. "So we stepped up the
poster walks and the picketing. All those businesses on Washington
and Jefferson, the department stores, the five-and-dime stores, wherever they had lunch counters, water fountains, these were the places
we targeted."[39]

Spring Arrives; Huntsville Heats Up

As winter turned to spring with no change from the white community, morale among sit-in organizers began to sag. The school system
was putting pressure on black teachers who might have wanted to participate in protests, and a number of black businessmen who had initially supported the movement dropped out because their white customers were abandoning them. To keep people's spirits up, the CSC
would invite in veteran speakers from earlier civil rights struggles, and
after a while people started kicking around the idea: why not invite
Martin Luther King Jr. up here? Hank Thomas remained in close touch
with the staff of the Southern Christian Leadership Conference (SCLC)
through this period, so they were aware and concerned about the flag-
ging enthusiasm level in Huntsville.[40] The SCLC agreed that a visit by
Dr. King might give the protest a needed shot in the arm. On March
19, 1962, twenty-five CSC members traveled out to the Huntsville airport, where they put Dr. King and Ralph Abernathy in Dr. Hereford's
brand-new yellow Cadillac convertible for a slow drive through town.
Next, they drove past the A&M campus so that the students could run
out and see the great civil rights leader. That afternoon King spoke to
about three hundred people at the First Missionary Baptist Church.
Later, at Oakwood College, more than two thousand people came out

to see King. But the news blackout continued. "His visit didn't even make the front page of the paper, which hurt us in our ability to bring outside pressure on the city," Hereford wrote. "We'd try get UPI and the AP interested in what we were doing, and the first thing they ask us was 'Did your local newspaper carry a story?' And we'd say, 'Well, no, not yet.' And they'd say, 'Well there's your answer. If the local paper doesn't carry it, why should we?'"[41]

With frustration mounting, Dr. Cashin, who had formed the CSC's psychological warfare committee, came up with a new idea. He and Hereford sat down for a talk with their wives, and the four of them hatched a scheme that everyone thought would finally break the blackout. "Jail 6, Pregnant Women 90 Days for Sitting-In" blared the April 24 edition of the national *Daily Defender*. "A pregnant woman, another woman who has a four-month-old baby, and three ministers were among seven persons sentenced to 90 days in jail here for sitting in at a lunch counter," the paper said, noting that the protest happened "in the city which calls itself 'the space capital of the world.'" The pregnant woman was listed as "Mrs. S. W. Hereford, III, wife of a physician." The woman holding the four-month-old was "Mrs. John Cashin, Jr., wife of a dentist." The group walked into Walgreens, the paper said, and "instead of hamburgers and coffee, were served with warrants."[42] Hereford said, "You talk about people rallying to the cause! Once the AP, UPI, *Jet* magazine, and others picked up the story, the local paper had to print it. It was embarrassed not to print it—a doctor's wife, expecting, and a dentist's wife, with a baby in her arms, carried off to jail. Yes, this story did go out."[43]

Now that the sit-in organizers had "the ball rolling," they had another trick up their sleeves.[44] "As you know, black people like to dress," Richard Hall said. "So, at Easter everybody would go and buy an outfit generally, if they could afford it."[45] In fact, according to Dr. Hereford, the Easter clothing splurge was the largest purchase most black Huntsvillians made all year (the second largest being for Christmas toys). On a visit to Nashville in the middle of the Huntsville protests, Hereford learned about a protest called "Blue Jean Easter" where African Americans, "instead of buying $100 suits and $100 dresses, they decided to spend five dollars on a pair of blue jeans for Easter, and I brought that idea back to Huntsville."[46] That Easter, Richard Hall said, African Americans in Huntsville "decided that they weren't going to buy anything from the merchants." The economic toll downtown was enormous. "There were twenty thousand black people in Madison County," Hereford said, "and ten thousand in the city, and if there are even ten thousand black people failing to buy $90 or $100 Easter outfits, that's a lot of money and losses for the merchants downtown. It could cost them a million dollars or more." As an extra, added dig at the storeowners, Hereford said, people did not even buy their blue jeans in Huntsville. "We bought them in Fayetteville, we bought them in Athens, we bought them in Decatur. We didn't even want the local merchants to have the five dollars."[47] When several hundred A&M students wore blue jeans in the segregated Huntsville Easter Parade to protest "the Negro's inferior position in the South," the black press made sure the Huntsville Blue Jean Easter was national news.[48]

The Blue Jean Easter's financial toll finally got Mayor Searcy to capitulate on the idea of a biracial committee, though he insisted on handpicking all the black members. The protesters argued that all the mayor's picks were "Uncle Toms," and insisted on their own people. The mayor said that if he used their people, he would not be able to find any whites to serve.[49] The back and forth only emboldened the protesters.

Buoyed by all the attention, they next decided to take their campaign to the very belly of the beast. In 1962, George Wallace was in a runoff election for governor against a young state senator named Ryan DeGraffenried, whom Wallace tried to paint as being in the back pocket of outsiders who opposed segregation.[50] On May 19, the Redstone Arsenal held an open house for employees' families. Management expected more than fifty thousand people to pour through the front gates, an irresistible opportunity for any politician in an election year.[51] While Wallace ranted to NASA and army employees, first in nearby Hartselle and then at the Huntsville courthouse, about standing "between you and those who would impose on you doctrines foreign to our way of life," the Huntsville protesters waited nearby.[52] As Wallace vowed to "fight for segregation in Alabama," they released hundreds of helium-filled balloons with messages written on the side. One said, "This balloon was released by Negro students in the courthouse yard at Huntsville, Alabama May 19, 1962. In this city millions of tax dollars are spent each day to build up Free World defenses, while city leaders, who benefit from these expenditures, oppose Free World policies."[53]

"Let's Talk This Over"

Throughout this period, the protestors were also making sure that the recipients of those tax dollars—the U.S. Army and NASA—knew their demands. In June, a series of constant "poster walks" in front of the MSFC's gates had a significant impact.[54] Attorney General Robert Kennedy, "the man carrying the ball most often in the government's civil rights push," was keeping a close eye on Alabama.[55] On June 5, Bart Slattery told Wernher von Braun that "word by the grapevine" had it that Kennedy was "eager to visit Huntsville." Slattery said that he "would view his visit with mixed emotions. We have a little picketing activity going on in Huntsville at the present time," he said, "and there is no telling what a visit from him might generate in the way of demonstrations, editorial comment and particularly rumors that he might be looking into our employment practices." Von Braun, knowing that NASA was still at the bottom of the list among government agencies on hiring blacks, understood that the rumors were not idle.[56] Slattery knew Kennedy would not come down without an official excuse. "Even in his position," Slattery said, "he has to have a reason for coming, especially to some kind of legal convention, FBI gathering, or some other affair which has some connection with his office." He told von Braun that he was looking "at other cities in the area for something that might give him an excuse for coming down." In other words, he was finding another city for Kennedy to go to, so that he would not end up visiting Huntsville. Slattery told von Braun all this in a memo, at the bottom of which the director underlined Slattery's name and wrote, "Let's talk this over."[57] Robert Kennedy did not visit Huntsville during the sit-ins.

As it turned out, the protesters did not need his attention to take their message truly national. Later in the month Dr. and Mrs. Hereford, along with Reverend Bell and an A&M professor, got in a car and drove north to Chicago. "We met two former Huntsville residents there, and the next morning we picketed in front of the Midwest Stock Exchange," Hereford said. They passed out leaflets that said, "Do not invest in companies that do business in Huntsville, Alabama." He continued, "And we talked to anyone who would listen about the companies that had government contracts on Redstone Arsenal and the space community." The Cashins, meanwhile, took a half-dozen A&M students and traveled to New York, where they "showed up outside the New York Stock Exchange. They started passing handbills that said, "To invest in Huntsville, Alabama is to invest in segregation" and "to bring in new plants and businesses to Huntsville aids segregation and subjects additional employees to racism." The handbills reported that great numbers of black people had been arrested since the sit-ins began and that city officials were still refusing to negotiate with the black community. The story went out on the AP and quickly got the attention of a city whose financial well-being depended on outside investment."[58]

This appears to have finally done the trick. "Not long afterward, city leaders decided they would have a trial desegregation of the lunch counters in Walgreens, Liggett's, Woolworths, and several other places in town."[59] With no riots and with physical harm, intimidation, and terror kept at a level that gave the appearance of normality, Huntsville desegregated. Almost simultaneously, Huntsville integrated the pediatric ward at the hospital—with the persuasion of Dr. Hereford, who was the CSC's one-man committee on desegregating hospitals—and by the end of the year had integrated the rest of the hospital as well.[60]

Huntsville's white population has largely controlled the narrative of this era. Consequently, credit for the city's ability to avoid widespread violence usually goes to Huntsville's white businessmen and their response to the pressure they felt from NASA. This clean, comprehensible, and self-serving way of looking at what happened is understandable if not entirely honest. In fact, the black community dragged Huntsville into desegregation kicking and screaming as hard and as loudly as any other community in the South. And while the sit-ins brought some changes, Huntsville remained far from normal, if normal meant all its citizens had an equal shot at the American Dream.

Why They Stayed Home

During the Huntsville sit-ins, as with Julius Montgomery, as with Frank Crossley, as with George Carruthers, Richard Hall and Delano

Hyder stayed away from the action. According to Sonnie Hereford, the sit-ins were "one of the most important civil rights victories in Alabama since the bus boycott in Montgomery"; yet all the time unrest was roiling Huntsville, Hall and Hyder stayed out of the way.[61] Is it right to fault them for this? Why was it that Hereford and Cashin had no problem picking up signs and marching in front of restaurants, meeting with the mayor, or shouting outside a George Wallace rally while Hyder and Hall were content to show up for work every day at NASA and let others carry the weight? In considering the answer, there are a number of elements to contemplate. First, Hereford and Cashin aside, the sit-ins in Huntsville were a young person's movement. The young had an assuredness about their own rightness, and that of their cause, that allowed them to lay doubts aside when it was time for action. This dynamic would appear again later in the 1960s when, as Theodis Ray noted, the nonviolent code of Martin Luther King Jr. gave way to the Black Panthers' shouts of "Burn, Baby, Burn."[62] There were certainly many older African Americans in Huntsville who saw the protests, polite and orderly though they were, as unseemly and potentially dangerous. Another consideration is that young people had a lot less to lose. According to Mary Ann Moore, who participated as a young woman in the Birmingham protests, "During that time most of our parents were afraid. The rationale was that when their bosses learned that they were participating with the Movement, then they were fired. In many cases the parents would encourage the children to go to the Movement."[63] Cashin and Hereford were doctors; they were self-employed and worked only in the black community, so they knew they would face no economic repercussions. In that regard, it is also worth remembering that as African American NASA employees, Hall and Hyder probably felt themselves more vulnerable than a white man like Marshall Keith. Black NASA workers in Alabama learned through the course of several discrimination grievance hearings that, in a pinch, the Marshall Space Flight Center personnel office was not their friend.[64]

A final reason Hall and Hyder felt comfortable and even obliged to remove themselves from Huntsville's unrest in 1962 was because, for them, Vice President Lyndon Johnson's plan to create equality through prosperity was working. In the 1950s, their experience and that of their parents and other relatives told them without question: there are just certain jobs that are not for us. There were doctors and nurses in their communities, but, according to James Jennings, who worked with Hall and Hyder, "the biggest aspiration was being a teacher." Hyder agreed that "preachers or teachers" were the top jobs blacks could hold. "The best would be a preacher or teacher," Richard Hall concurred. "There was no gray area."[65] Those limited options were of course only available to African Americans who had an education. Without college, the best jobs were at the low end of agriculture—tractor driver or sharecropper. As Sonnie Hereford remembered, in Huntsville, "Of the black people I knew, probably 80% were farmers. A few worked for some of the new and used car companies, a few on a Coca-Cola truck, a few on the furniture truck. A few were ministers and then some schoolteachers, but nobody had any real technical jobs, at least no black people I was acquainted with. Several worked menial jobs in the textile mills."[66] Knowing this, the parents of these NASA workers pushed hard for them to stay in school. They knew, as Harry T. Moore had learned from those who went before him, that if you were black in America, life was tough—the playing field was not level, the odds were against you. But if a change was coming, the only hope for advancement was to get an education. Delano Hyder's story is typical. "My father worked for the coal company," he said, "and when he would come home in the evening, he was so covered with soot 'til I knew I didn't want to be that. And he told me I could be anything I want to be." Parents, teachers, and other leaders in the community all pushed these young people to go to college because, as Richard Hall put it, "I knew that if I go to school, I'll never have to plow that mule or pick that cotton."[67] Johnson sought to locate the space program in the South in order to create new jobs for African Americans to aspire to and then hold. Hyder and Hall were beneficiaries of that. At the time of the sit-ins, the unmerciful restrictions placed on African Americans looking for jobs in math, science, and engineering were beginning to lift, thanks to the space program. Federal salaries were (as they are today) based on an employee's "General Schedule," or GS level. A physical science aide like Richard Hall was a GS-3, making $3,175 a year. Although that was the same as a janitor or a low-level sales clerk, for a black man in the early 1960s in Alabama, it was a living wage with the important benefit of job security.[68]

Not Qualified

Despite the appeal of these jobs, NASA—try as it might—could not seem to hire a sufficient number of African Americans to keep its hiring stats competitive with other federal agencies. NASA claimed its greatest problem with equal employment was that it could not find blacks to hire. It said hiring practices at contractors often belied their declarations of "equal employment opportunity" and that prospective black employees were likely to view an all-white workforce as more indicative of corporate hiring practices than any self-promoting advertisements. Second, the agency said black schools were located in places less accessible to recruiters than white colleges, making contact difficult. Additionally, recruiting firms and the agency maintained standards and requirements, such as qualifying examinations, beyond the black applicants' training.[69] As a result, the agency continued to search and was continually frustrated.

NASA first set out hoping to recruit blacks from its local communities, but in Huntsville it quickly became clear that the local black colleges were not going to be turning out NASA-ready engineers any time soon. Several contractors had promised to assist the schools "in improving their facilities, curriculum and faculties," but the reality was that the schools, A&M in particular, just were not ready.[70] A&M had no library and repeatedly lost its accreditation.[71] NASA tried to assist the school in applying for grants, but the work never progressed—and not because of any failing on the agency's part. The school was not interested and they did not pursue the money.[72] When NASA looked locally for African American physicists, it found A&M did not possess the right test equipment in its lab.[73] And it had no aerospace engineering courses—only electrical, mechanical, and civil engineering.[74]

Technology jobs were not the only work NASA could provide. NASA and the space program required clerical staff, accountants, personnel systems, computer, and data processing staff. There were contractors around the country that were successful in making "a distinct impact" on their goals of workplace diversity by hiring African Americans with these skills.[75] NASA, however, claimed to have comparable problems finding African Americans in those disciplines, too. A 1963 report to headquarters by MSFC talked about the center's attempts to recruit African American office workers in the Huntsville area. The center was unable "to locate an adequate supply" of qualified African American clerk-typists or "professional and semiprofessional" employees, the report said, mentioning that of 475 applicants for clerk-typist positions at the Redstone Arsenal, only 9 "were identified as being minority group applicants," and none of those nine were qualified.[76]

Once the agency realized it could not recruit locally, it decided to strike out around the country and try. Its first effort was a dismal failure. NASA sent a recruiter to the Tuskegee Institute in Alabama, but only a few of the graduating engineers would talk to him. The recruiter's explanation was that "all the rest said they were going north," but Richard Hall had a more realistic explanation. NASA, he said, would only send out recruiters who were white. "It would be a difficult problem for black recruiters to talk to a person about coming to Huntsville. They sent out white recruiters, and they end up with basically zero."

With all that, however, it is noteworthy that Hyder and Hall had no trouble finding jobs with NASA, and that it had no trouble finding them. Both men graduated from A&M in 1957. A friend told Hyder about a job he held in the computation lab at the Redstone Arsenal, calculating the thrust of missiles. Hyder had majored in chemistry with a minor in math and biology, so this kind of work was right up his alley. As he recalled, "I just got applications around the middle of the month and then a month later after I put the applications in, I got a call for an interview." He described the process of getting a job at the Redstone Arsenal as "kind of simple and easy. It was less than a month at least from the time I graduated to the time I went to work."[77] Hall graduated from A&M with a degree in industrial arts and a minor in mathematics. "I went to the post office—at that time, they posted their jobs on the bulletin board at the post office—and I saw an application out there for employees. I filled it out and send it to Atlanta, which was the 3rd Army area."[78] NASA hired him as a physical science aide within a matter of days.

The nation's 106 black colleges were certainly graduating hundreds of people like Hyder and Hall who had skills that enabled them to do the kind of work NASA needed. Hyder and Hall, for instance, both started as film readers. In a time before technicians could read gauges remotely, engineers would place a steel-covered 16-mm movie camera inside the belly of a missile and it would shoot a movie of the gauges during the test firing. After the film was developed, someone would read it and write down the numbers that appeared on the gauge. They would hand that off to someone else to do the math. The job required some math skills and a tolerance for tedium. Any number of African American college graduates might have taken those jobs. But the deeply engrained reputation of the South for violence and discrimination was a high hurdle to get over. It is also fair to ask whether NASA's people in the South necessarily wanted to get over the hurdle, which brings us back to the question of how NASA's white staff in Huntsville felt about the important issue of racial equality. According to James Jennings, Hyder and Hall, along with the other African Americans working in the computer lab when he got to NASA, did not come into the government as full professionals. Despite having "graduated from school with full degrees in the technical areas—math, chemistry, some of the sciences—they were actually brought into the government as technicians."[79] Their advancement was blocked—they were not able to get better jobs—for one simple reason. While NASA had classes that would have allowed them to advance, Richard Hall said, "They had a place for the whites to go and get their training, but there was no place for blacks to go and get theirs."[80] The battle to change that situation would prove to be a severe test of whether the agency and its people were committed to change.

The Schoolhouse Door in Huntsville

When George Wallace lost the governorship to John Patterson in 1958, he firmly believed the loss was due to his opponent taking the stronger position on segregation. When Wallace ran in 1962, he was not going to make that mistake again. No politician in Alabama was going to be a bigger white supremacist than George Wallace was, and he was going to make sure that when someone from the outside challenged the racists of Alabama, they knew who had their back. "Let us send this message back to Washington," Wallace said at his inauguration. "We are standing up, and the heel of tyranny does not fit the neck of an upright man." The rules in Alabama were clear, he said. African Americans belonged in their "separate racial station." His message to anyone wanting to alter that station and bring "the false doctrine of communistic amalgamation" was unambiguous: "We will not surrender our system of government, our freedom of race and religion. That freedom was won at a hard price," Wallace said, "and if it requires a hard price to retain it, we are able and quite willing to pay it."[81]

There was a hint of the bigger fight to come in the days just after Wallace's election, when the University of Alabama's Huntsville Center (UA-HC, the predecessor to the University of Alabama–Huntsville) announced an extension seminar on current affairs. The six-week course was just the sort of community enrichment program Huntsville loved to bask in as evidence of how different and modern it was. Speakers would come in from around the region to engage lifelong learners with their reflections on domestic issues, the upcoming session of Congress, relations with Red China, the new nations of Africa, and the standoff in Berlin. The center offered classes like these alongside the posting for advanced training at MSFC and the Redstone Arsenal. Employees and their spouses enrolled to gain advanced skills and knowledge. After four months of planning, all was ready when, on February 2, the center abruptly announced that the extension seminar had been canceled. Those who had enrolled in the class surely met this news with disappointment, but also relief. While it was sad that the school had canceled this lovely opportunity to learn and grow, the alternative would have been unthinkable under the circumstances. The university canceled the class because a black woman had signed up to attend. And not just any black woman. Joan Cashin, the wife of the dentist who had formed the psychological warfare unit of the Huntsville Community Service Committee—the woman who got herself arrested for sitting-in at Walgreens with her four-month-old baby and her pregnant friend.[82]

For the same reason UA-HC would not allow Richard Hall and Delano Hyder to take classes that might help them advance, Joan Cashin was not going to learn about current events at a whites-only college. As Wallace had said, blacks belonged within their "separate racial station." Those were the rules in Alabama. The university said it canceled the class because no speakers were available, but an article in the black *Atlanta Daily World* pointed out that that claim was transparently false. Just recently, the paper reported, officials had "stopped all admissions to the University at Tuscaloosa for next semester after Negroes applied there."[83] There had been no African Americans at any branch of the university since the breathtakingly short matriculation exactly seven years earlier of a woman named Autherine Lucy, "the first Negro actually enrolled in a state university in the 'hard-core' [segregation] states."[84] On February 3, 1956, according to a contemporaneous account in the *Journal of Negro Education*, Ms. Lucy was admitted to classes at the University of Alabama "after all manner of fulminating legal actions on the part of the University failed."[85] For the next three days, "the University of Alabama acted so as to appear a mob-respecting and law-rejecting institution" as rioters and demonstrators swept across the campus and the board of trustees banned Lucy "until further notice." She went back to the judge who had supported her admission in the first place and he reinstated her. But three weeks later, the board of trustees expelled her for making "accusations against University officials which were slanderous and groundless."[86]

The "Lucy incident" was infamous among African Americans in Alabama, so Mrs. Cashin clearly knew what she was doing when she went over to apply for the extension seminar. She arranged for a white friend to go over to campus with her. Mrs. Cashin went in first and enrolled. Then she walked out and her white friend walked in and applied. This way, as the *Atlanta Daily World* said, "university officials would not be able to say Mrs. Cashin was refused admission because she applied too late."[87] The CSC psychological warfare committee had struck again, outfoxing the UA-HC administration. Denied their go-to excuse, the school simply decided to cancel the whole thing.

Alabama Desegregation May Start at Huntsville

Mrs. Cashin's episode was just a warm-up, as it turned out. Wallace vowed to send a "message back to Washington" during his inaugural address, and in 1963 it appeared NASA would be his chosen messenger. "Alabama Desegregation May Start at Huntsville" was the headline of a story in the March 25 edition of the *Washington Post*. "The desegregation showdown in Alabama apparently will come in June in this Space Age city, rather than in September elsewhere in the state as had been expected," the article said. Two black men, Marvin Phillips Carroll, a twenty-seven-year-old electronics engineer at Huntsville's Army Ballistic Missile Agency, and David Mack McGlathery, a twenty-six-year-old mathematician in the Research Projects Division of the Nuclear and Ion Physics Branch at MSFC, had applied to take classes at the University of Alabama–Huntsville Center.

Like Hall and Hyder, McGlathery had a bachelor's degree from Alabama A&M. His was in math and while he was at school his excellent grades earned him the nickname "Mr. Brains."[88] He had applied to the university—as Joan Cashin had—under the program to allow NASA employees to further their professional qualifications. Rejected by university officials in March, McGlathery and Carroll planned to "press immediately in federal court for admission to the summer quarter in June."[89] That this was happening in a NASA community and to a NASA employee was potential trouble enough for the agency. But in addition, NASA had recently given the University of Alabama $600,000 to start a space research institute in Huntsville, and officials at the agency and the college were concerned the governor might impose on the legislature to block that money.[90] That possibility got NASA's attention and threatened to escalate the situation into a full-blown crisis.

Unlike Julius Montgomery's spur-of-the-moment decision "for the better good of everybody" to pull out of the first class at Brevard Engineering College back in 1959, this decision was heavily freighted with local, regional, national, and international symbolism. Although he denies it, McGlathery told William Ellis of the *Nation* that an official at MSFC convinced him to avoid controversy and withdraw his application for his own good and the good of the space program. Ellis wrote that the official, identified in the *Washington Post* as "a second-in-command to Wernher von Braun" named Ernst Stuhlinger, "advised him that while he was qualified and certainly entitled to take courses at the university, perhaps he wouldn't care to become the center of what could develop into a loud and nasty controversy. Wouldn't it be better to wait awhile?"[91] The forces aligned around the unassuming mathematician were immense. On one side were those of Wallace's "Heart of the Great Anglo-Saxon Southland." McGlathery received death threats and prank phone calls.[92] On another side was the U.S. Justice Department; the U.S. attorney general in charge of civil rights, Burke Marshall, spoke with McGlathery and Carroll in Wernher von Braun's office.[93] Huge monetary investments were at stake. The integration of Ole Miss and its symbolic importance was fresh in everyone's mind, as were the riots that followed. If the same thing happened in Huntsville, it could do real damage to the space program and might mean America would not land a man on the Moon by the end of the decade. In the moment, the decision was agonizing.[94]

McGlathery's comments at the time, however, fall in line with attitudes typical of Hall and Hyder—principally a desire not to rock the boat.[95] "I didn't want to be in a position for someone to say that I was responsible for scuttling the space research institute," McGlathery told Ellis. He said the same thing to newspapers in the moment, including telling the *Washington Post* that he specifically did not want to create a situation like the one James Meredith had created in Mississippi. Richard Hall, who had "known Dave since he was a kid," said McGlathery's decision was really a no-brainer—it was the same decision he or anyone else would have made. "It's just an event that happened I guess," Hall said. "I don't have any strong feelings one way or the other about that."[96]

George Wallace felt otherwise. The governor attended a UA board of trustees meeting where they discussed McGlathery and Carroll. The governor reiterated his strong opposition to university integration and announced his intention to call Alabama's U.S. senators, Lister Hill and John Sparkman, to ask their help in getting McGlathery and Carroll transferred out of their jobs. Burke Marshall told Attorney General Kennedy that he expected Wallace to begin background investigations of the two NASA employees. Furthermore, "the governor also said that he intended to call Werner [*sic*] von Braun and tell him that an incident in Huntsville would be bad for the space program."[97]

NASA and the nation would come to fully grasp just what such an "incident" might be like when, beginning around April 15, unrest brought the community of Birmingham "to the brink of racial war."[98] By April 19, 1963, Birmingham was in turmoil. In a few hours, images of white police officers attacking black children with dogs and water cannon would stun the nation. Police had just arrested Martin Luther King Jr. for seeking, as he would say in his famous "Letter from a Birmingham Jail," "to create such a crisis and foster such a tension that a community which has constantly refused to negotiate is forced to confront the issue."[99] That day, NASA administrator James Webb sent a letter to the heads of his major centers in the South. In it, the administrator, who had built his agency's reputation on never deviating from its mission, was asking his men to take their eye off the ball.

"Time-Consuming, Difficult to Implement, and Full of Intangibles"

"As managers of government activities," Webb wrote, "we must assume responsibilities for implementing many programs which may seem not to contribute immediately and directly to our program schedules." We take these responsibilities on, he said, when Congress and the president demand it "for the good of the country," even though "sometimes, these are time-consuming, difficult to implement, and full of intangibles." Unrest like that in Birmingham was behind Webb's letter, and now, Webb told the managers, "assuring equal employment opportunity, regardless of race, color, creed or national origin" was their struggle, too. "The problems you face in this area are somewhat more difficult than they may be in other parts of the country. This makes it doubly important that supervisors at all levels, and recruiting officers, in particular, be fully informed as to their responsibilities for assuring equal employment opportunity."[100]

It is not clear what prompted Webb to pick this time to send the letter. While King's trial was on, the Justice Department was expressing the opinion that "the Birmingham situation will not become as protracted or as bitter" as past civil rights–related unrest. Up to that point it had not been.[101] Webb was not responding to the now iconic images of Birmingham's police brutality, because the nation at large had not seen them yet. While Webb would certainly have his arm twisted by the vice president and the attorney general as 1963 went on, there is no documentary evidence that that effort had begun in April. In fact, even though Webb's letter to von Braun acknowledged that the Marshall Space Flight Center's equal employment efforts "have been less successful than those of others" in the agency, he said that the chairman of the Civil Service Commission had told the vice president that progress on hiring African Americans at NASA overall "was especially noteworthy."[102] So why did he send it?

Webb attached the letter to a report that was "a composite of activity throughout NASA" on equal employment, so that may have been reason enough. But the letter was lengthy—at least the version that went to von Braun and Kurt Debus at Cape Canaveral. It was much more than just a cover note on a report. Webb did take his role as a member of the President's Committee on Equal Employment Opportunity seriously. He attended meetings with a frequency that seems unusual for the administrator of an agency that had nothing directly to do with labor, justice, or any of the other burning issues of civil rights activity. It is also true that he, like Vice President Johnson, believed the federal government had a role in changing the society of the South and often acknowledged his understanding of NASA's place in that plan. Also, like many New Deal liberals, Webb passionately believed in the government's ability to solve problems. On that score, he also believed in a particular type of Magical Thinking, which he called "Space Age management." Webb saw in the space program "the seeds of transformation for the nation as a whole." Space Age management, Webb thought, represented "a panacea for all of the ills of society."[103] As Webb put it at the Space, Science, and Urban Life conference in Oakland,

> Advances in science and technology are largely responsible for this thrust of urbanization. It seems appropriate therefore that science and technology should increasingly seek ways and means through which some of its efforts may be turned in increasing amounts to helping solve the problems, urban and rural, which these changes are bringing about.[104]

It was always difficult to nail Webb down on precisely what the term "Space Age management" meant aside from creating organizational structures that could adapt to changing circumstances and "the desire to use the power of the federal government to be active about improving the lives of Americans." But he did believe that the way to solve big problems was by thinking big and acting big.[105] Perhaps Webb thought that NASA's vastness would have spillover effects on civil rights. He certainly saw it as a place that could exert its influence over the hiring practices in the South. Now his agency had to act to solve this particular problem and others that spun off from it.

As the violence in Birmingham galvanized the nation, David McGlathery and Marvin Carroll moved ahead with their plan to integrate the Huntsville university center. This time no one at NASA tried to stop them. They filed suit in federal court on May 9. To its credit, the UA board did not challenge the decision, but Wallace had other plans. A few days later, NASA headquarters contacted the Marshall Space Flight Center with an urgent warning: an episode that would become one of most important flashpoints of the civil rights movement was shaping up to happen in Huntsville.

Headquarters had just received "a book from the Department of Justice" with a memo attached to it, signed by Attorney General Robert Kennedy. The book contained a list of companies with major installations in Alabama and the names and phone numbers of their chief executives. On orders from the Kennedy brothers, officials in the president's cabinet were to call all the CEOs they knew and ask them to contact Wallace and get him to stop a major action he was planning. The University of Alabama was going to be required to "admit at least two Negroes the week of June 10, one to the graduate Center at Huntsville," Kennedy's memo said. "Unless he changes direction, Governor Wallace intends to create a major incident" by coming personally to the University of Alabama–Huntsville Center and standing in the schoolhouse door to block David McGlathery and Marvin Carroll from enrolling.[106] In a reference to the riots in Mississippi, Kennedy said that if this were to happen, it would create "a severe problem in avoiding violence of the sort that occurred at Oxford."[107] Everything NASA feared when it asked McGlathery back in March to withdraw now looked like it would happen. Huntsville was poised to erupt like another Birmingham, where, by mid-month, President Kennedy had decided to send three thousand soldiers.

A few days later, the Justice Department circulated yet another large book of CEO names. Wallace now planned, in addition to personally blocking integration in Huntsville, to stand at the university's main campus at Tuscaloosa. The Kennedy administration, for its part, once again appealed to Wallace's practical side by having his friends talk him out of it. The department's book, compiled with the quiet assistance of a University of Alabama trustee, had the names of major businesses in Alabama and their officers.[108] Again, the White House sent the list of names out to every member of the president's cabinet, and this time to the secretaries of the army, navy, and air force, the Securities and Exchange Commission chairman, the chairman of the Federal Communications Commission, and the administrators of the Federal Aviation Administration and NASA. The list had a memo attached to it asking the government officials "to indicate those persons known personally to them and to return the notebook and this information to the Department of Justice." After those lists were compiled, they were sent back to the cabinet officers and others "with a memorandum requesting the recipients to telephone the persons on their list and to urge these persons to contact Governor Wallace and try to persuade him to follow a course of moderation so that Alabama would not suffer the harm that came to Mississippi." Among the firms contacted by NASA were Chrysler, Coca-Cola Bottling, General Electric, IBM, Scott Paper, and Thiokol Chemical.[109]

Thomas McCabe of Scott Paper responded positively to the call he received from James Webb, and, in addition to contacting Wallace, instituted limited integration in his plant. Although sympathetic to the integration cause, McCabe worried about extremists on both sides of the issue and was convinced that Wallace would maintain a hard line on race. Indicative of McCabe's concern was his request that Webb keep the information he provided confidential.[110] J. A. Barclay, manager of a division of Northrop Corporation, also contacted Wallace. Speaking as "a resident of Alabama, by choice," Barclay urged the governor to "produce a climate which will distinguish Alabama as being as progressive in the social field as it is now known to be in the technical and space fields." Foreshadowing a problem that would loom larger for NASA in Alabama as the civil rights crisis progressed, he reminded the governor that "one of the tasks which we in the aerospace industry must face in building sizable operations in Huntsville is to lure experienced people from their present location," a task that was becoming increasingly difficult. He related the experience of "two engineers with graduate degrees who agreed to move to Huntsville [but then] changed their minds," and told the governor that the engineers informed him that they had seen the riots and the unrest in Mississippi and worried that a comparable "racial mess" might occur in Alabama. "You are certainly entitled to your opinions," he told Wallace, but he urged him to allow the UA integration to be "handled with the laws of our nation and with the dignity and wisdom which the citizens of Alabama have a right to expect from their chief executive."[111]

NASA's role in pressuring Wallace is commendable. However, while the agency's people made phone calls asking others to contact Wallace, NASA itself passed up its one opportunity to confront the governor directly during this period. MSFC PR director Bart Slattery convinced Wernher von Braun not to attend the groundbreaking for a new Lockheed facility inside the Huntsville Research Park specifically "because

Governor Wallace will be there."[112] Writing to a Harvard undergraduate about Wallace's stand in the schoolhouse door a few days after the fact, von Braun said, "While under certain circumstances it is unwise for federal employees to interject themselves into state matters, there frequently comes a time when one must take a position. Such an occasion presented itself immediately before the registration of the three Negro students."[113] While it did, it is a shame von Braun wasted it. If there was one thing the world knew Wernher von Braun for, it was delivering a payload to its target with devastating effectiveness.

By the end of May 1963, it was still not clear whether Wallace would take his stand in Huntsville or Tuscaloosa. In addition to McGlathery, Vivian Malone and Jimmy Hood had filed suit in April to attend classes at the university's main campus.[114] Marvin Carroll, the army electronics engineer, was out of the picture at this point. The investigations Wallace had ordered back in March had revealed that as a teen Carroll had fathered several children out of wedlock and that a court had charged him with the crime of "bastardy" and given him a twelve-month suspended sentence. On his application for admission to UA-HC, Carroll checked "no" in the box asking, "Have you ever been officially charged with a criminal offense other than a minor traffic violation?" Discovery of his false statement gave the university justification to dismiss his application.[115] Justice Department officials recommended that troops be readied at the Redstone Arsenal in case the confrontation came in Huntsville. But as May turned to June, it became clear that Wallace was not looking to turn Alabama into another Mississippi. The university's director of public relations told Justice Department officials on June 3 that Wallace planned to "make a gesture, then step aside."[116] That is precisely what he did on June 11, once deputy U.S. attorney general Nicholas Katzenbach told him it was his "sad duty to ask you to step aside under the orders of the President of the United States."[117]

The following day, as reported in the *Atlanta Daily World*, David McGlathery enrolled at the University of Alabama–Huntsville Center, "without attracting a single white spectator." That description was not technically true. As the article later pointed out, "65 news men and about a dozen helmeted police officers" watched McGlathery's enrollment. The paper said he arrived at the campus "alone, parked his car in the student parking lot and walked into the building." McGlathery made a statement to the waiting press after he had registered. He told them, "This represents a new challenge for me. It is up to me to make good at it, not necessarily for the sake of my race, but for myself."[118] As it turned out, McGlathery failed his math course that summer, telling William Ellis of the *Nation* that it "was too tough for me." He re-entered the university in January 1964, posting an "A" in math and becoming an unlikely hero of the civil rights movement.[119]

"I Was So Busy During That Time"

It should not be surprising to learn that Richard Hall treated all the excitement surrounding Huntsville in 1962 and 1963 with a shrug. His reason was much like that of George Carruthers in Washington, DC. Though he did not claim that what happened did not actually happen, he did say that he was too busy to notice. "I was in testing of engines" in those years, he said, "and I was so busy during that time—I was working like 12 hours a day and eight hours on Saturday." With all that work, "you stay so busy [that] you don't have time to think about all those other things going on. You just try to solve your own little problems."[120] As for the day-to-day discrimination, the inability to advance professionally, the governor blocking progress, and NASA not stepping up, Delano Hyder was similarly nonplussed. "I didn't have any hard feelings against anybody about it," he said. He was just happy that circumstances and timing allowed him to have the opportunity to take part in what he considers one of humankind's greatest achievements. "I was excited myself," he said about the space program. "Because this is to be something that I can tell my grandkids about, that I worked for NASA and that we would be part of something that's going to affect the future."[121] What he was proud to be part of was moving forward the future of humanity in space. Those other things—what went on in one Alabama or the other—none of that bothered him all that much.

On June 11, 1963, President Kennedy responded to Wallace by delivering a beautiful and eloquent nationally televised address on civil rights. The goings-on in Alabama did bother the president, finally. Historian Hugh Davis Graham wrote that until that night, "no president since Lincoln had addressed the nation in such forthright moral terms about American race relations."[122] The president also took a step he had avoided and had hoped to keep avoiding: announcing that within days he would introduce a civil rights bill to Congress. At the same time, he geared up the administration's existing civil rights machinery significantly. For the remainder of the year, along with the rest of the federal government, the administration coerced and cajoled NASA into doing everything it possibly could—and even things it could not—to support the new priority. One man at NASA took the president's action to heart, managing to address—at least on a small scale—much of what President Kennedy called for that night and make it a reality.

The Country Spartacus
Clyde Foster

I think I did some good, and I feel good about it. It introduced politics into their minds.

CLYDE FOSTER

The PCEEO had enforcement powers and conducted investigations, but, on a day-to-day basis, the president's policies for integrating the federal workforce were the province of people on the ground. As rules trickled down from Washington to the actual federal outposts in the states, those responsible for imposing them and those who had to comply met their arrival with a range of feelings, from enthusiasm to hostility, to varying degrees of both. At NASA, there were those who applied themselves to the task of integration and racial understanding both officially and unofficially. Some did it because it was their job, but others had their own agendas.

Among the African American pioneers of NASA, there is no one who engaged in more unofficial acts to promote compliance with the spirit of equal employment than Clyde Foster. During his nearly thirty years at NASA, Foster—operating without portfolio—got more black workers jobs, advanced the education of more black engineering students, and cleared the path toward equality more enduringly and forcefully than any other African American NASA worker. Even more remarkable, he did all of this through conventional channels. In what will be a familiar refrain by now, he never marched, he never protested, he never sued, he never threatened. To convey just a portion of the significance of Foster's nonviolent activism to the space agency and the greater cause of civil rights, consider that in 1962 the NASA advanced training program in Alabama was whites only, and the United States had only two black mayors and no black police chiefs.[1] Ten years later, thanks largely to this man, whom *Ebony* magazine dubbed "a country Spartacus," none of that would be true.[2]

"Two Sets of Everything"

Foster grew up in Birmingham, Alabama, which, by the dawn of the Space Age, had spent generations mired in Jim Crow. When Foster was growing up, he said, there were "two sets of everything, one for colored and one for the white. Signs was posted on water fountains, restrooms, and at that time, you could not purchase a hamburger or hotdog because there was no public facilities [that allowed blacks], including restaurants."[3] Police harassment was a constant threat when he was young. "Whenever they would see a group of black kids assembled together, it was always some reason to go after them." Because of that, and just in general, he said, "You were afraid to travel on the wrong side of town, or be seen in certain areas of the city." The whole experience growing up, he said, "was very, very hateful."[4]

From that experience, Foster could clearly sense the limits that society would place on a black man professionally and, with the nation in the depths of the Great Depression, he made plans to get out. "I told my father I am not going to follow into his footsteps," he said. "He would often come home and talk about how tired he was and I told him, I said, 'Dad, I'm going to do something different.'" He knew that college was the only way out, and so he left Birmingham for Huntsville, enrolling at Alabama A&M College in 1949. "I determined at a very early age that I wanted to further my education."

In 1949, Wernher von Braun and the German rocket scientists had also recently arrived in town, and von Braun was pushing hard for Alabama to improve education and hence the state's future. He addressed the state legislature to request a $3 million bond issue for a research institute, and shortly thereafter, the state funded his request when the University of Alabama opened a Huntsville center for white students in 1950. His speech to the legislature was famous in its time, but less well known and considerably less successful was a trip von Braun took to Alabama A&M in 1952.

There, he recruited Clyde Foster and a group of other natural science majors into a quixotic lobbying campaign at a local white high school. Foster was not really sure what von Braun aimed to accomplish, but he believed that the scientist was looking for a way to interest the area's white high school students ("people who had only ever worked on tractors," was how Clyde Foster remembered them) in the wonders of outer space and the exciting fields of science and engineering. Why he thought black students were the best way to accomplish this is lost to history, but the result was that Foster and six others found themselves standing on stage in front of an auditorium filled with some of the very kids who had thrown garbage out their school bus windows at Sonnie Hereford. Von Braun had not asked the black students to prepare anything; he had written a script about his plans for Alabama, about flying to the Moon and about the need for education, and had the black students stand on stage and read it aloud. Von Braun "developed the write-up and we just had to follow the reading," Foster recalled.[5] The reaction was what one might expect. "Nobody was listening," Foster said. Over the years, the great rocket scientist took extraordinary steps to improve race relations in Alabama and to encourage scientific learning at black colleges—extraordinary for the times and the place and extraordinary too because he had spent so many years building weapons for the twentieth century's most notorious racist, Adolf Hitler. The actions taken on race relations by the Marshall Space Flight Center would go above and beyond anything done elsewhere by NASA in the South, and von Braun actually paid attention to the issue and was as engaged in it as he could be with all of his other responsibilities.[6] Wernher von Braun was far from clueless when it came to race relations in Alabama. Perhaps this particularly astonishing lack of cultural sensitivity is attributable to von Braun still being—in 1952—a fish out of water. While the assembly was a disaster, it did have one positive result and that was to introduce Clyde Foster to the space program; as it turned out, he would have as profound an impact on the space program as it would have on him.

Back to Huntsville

Foster's road to NASA still had a turn or two. He left Huntsville for the Korean War right after school, but not before—as *Ebony* put it—marrying one of Huntsville's "attractive natives, now Mrs. Dorothy Foster."[7] After his discharge from the army, Foster heard from a col-

Clyde Foster, seen here in front of the Marshall Space Flight Center in 1965, got more black workers jobs, advanced the education of more black engineering students, and cleared the path toward equality more enduringly and forcefully than any other African American NASA employee. Photo courtesy of Don Rutledge.

Clyde Foster, newly appointed mayor of Triana, Alabama, talks with his police chief, one of the first African American police chiefs in Alabama. Foster discovered Triana's defunct charter in the Alabama State Archives and revived the town in 1964. Photo courtesy of Don Rutledge.

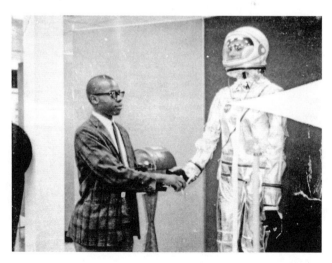

When Morgan Watson and six other students from Southern University–Baton Rouge became NASA's first African American engineers in 1964, Watson said, "We felt that the whole image of black people was riding on us as professionals and we could not fail." Family photograph courtesy of Morgan Watson.

George Bourda was one of six students from Southern University–Baton Rouge who in 1964 became NASA's first African American engineers in the South. "I look white," Bourda said, and, as a result, "they thought I was one of the guys. Every now and then they'd slip up and say something and [only then would] they realize I was there." Family photograph courtesy of George Bourda.

By 1965, Wernher von Braun, director of the Marshall Space Flight Center, was one of NASA's principal campaigners for civil rights in the South. Here, the former Nazi visits Miles College in Fairfield, Alabama, for the groundbreaking of a new science building. Photo courtesy of Miles College.

Alabama governor George Wallace (left), Marshall Space Flight Center director Wernher von Braun (center), and NASA administrator James Webb (right) meet at MSFC to witness the first test firing of a Saturn V Booster, along with members of the Alabama legislature and members of the press. NASA photo.

Morgan Watson, NASA's first African American engineer (left), and Julius Montgomery, the first African American hired as a professional at Cape Canaveral (second from left), speak on a panel at the National Air and Space Museum with astronauts Leland Melvin and Mae Jemison. This event was held in February 2010. Photo by Eric Long, Smithsonian National Air and Space Museum (NASM 2009-30619).

lege friend in Selma, Alabama, who mentioned that the school where he taught needed science teachers. Foster had majored in science and so he said, "I told my wife and I told my family that I was going to go down to Selma and apply for a teaching job. That was one of the things that really made my father happy," he remembered. "He was the most happy person in the world that he had a son—first graduate son, out of college—was going to become a teacher." Foster enjoyed the work in Selma. "I was very, very much delighted when I would look into the kids' eyes and see them grasp some of the things that I was teaching." His wife Dorothy was miserable, though. For a black Huntsvillian, living in Selma was a nightmare. Whites seemed to go out of their way to make you feel subhuman, Clyde Foster said, "moving about the city, and certainly when you would go into stores to be waited on. The individual would take their own time and turn around and ask you 'What you want?' in a very mean-spirited way. She cried every day she was in Selma," he said. He recalled, "She would say, 'I told you when we got married that I would follow you to the end of the earth. And then you carried me there!'"[8]

Foster's story of obtaining employment at NASA calls into question a lot of the conventional wisdom explaining why so few African Americans worked at the agency (throughout NASA's period of greatest growth, the number never got much beyond about 3 percent).[9] Another friend found a nontechnical job at the Army Ballistic Missile Agency (ABMA), a predecessor to NASA, "and he made mention that they were looking for technicians." When Foster wrote for and received an application, Dorothy could not have been happier. "She filled that dang thing out and sent it in for me," he remembered, laughing. When ABMA called him in for an interview, he said, "she shouted! She was ready to get out of there!" He got the job and Dorothy "got her cousin to come to Selma to load the two or three pieces of furniture we had and then to move us back to here."[10] That is a story clearly outside the realm of NASA's conventional wisdom about the roadblocks that kept African Americans from their doors.

Foster was one of the massive team of people who, during rocket test firings, calculated numbers from gauges set deep inside missiles and rocket engines. They kept track of the missile's speed and its heat and then reduced those numbers to averages that allowed engineers to calculate how much thrust a missile had, how much wind resistance it would meet, and whether the combination would allow it to move off into space, cranking out numbers on forty-pound steel-crank and gear-driven calculating machines. Within a few months of being there, he got his fraternity brother Delano Hyder a job in the computation lab.

They were both there when, in 1958, ABMA became a part of the new space agency. Shortly after that, the presidential administration turned over and President Kennedy was announcing that Americans would go to the Moon. Kennedy's announcement actually came at a time when Foster was trying to decide whether to quit NASA, but something the president mentioned in his speech changed his mind. "I elected to go to the space program," he said, "because Kennedy had made that famous speech: 'We're going to be on the Moon within ten years.' I said, 'Boy, I could have some longevity here!'" After he made his decision to stay, Foster talked it over with his father. "That's wise, son," the older man told him, "but I'll tell you what; you do all you can, but you are not going to the Moon."[11] That was a healthy dose of skepticism. Perhaps the years of pain, repression, and unremitting disappointment living life as a black man in early twentieth-century Alabama had prepared Foster's father to know better than most: the Space Age was no magical panacea. Working to bring racial equality to America was going to be a dream just as hard to achieve and maybe just as elusive as going to the Moon.

"A New Era of Equality According to Ability"

In 1962 it was not only the grandchildren of slaves seeking their first-ever piece of the American Dream who imagined that the Space Age might be linked to better race relations. The idea was actually a subject of scientific inquiry. That spring, NASA awarded an $181,000 grant to the American Academy of Arts and Sciences "for the purposes of studying, over a period of approximately 2 years, the relationships of space efforts to US society." The academy knew that the space program would be huge, would absorb the bulk of American scientific and engineering expertise, would fall into place quickly, and could therefore create "a diversity of unplanned effects on the organization of society."[12] It requested the money from NASA in order to understand those effects, including the impact of the space program on race relations. "While many have questioned the extent to which NASA's efforts are at the forefront of science and technology," the academy said, "there can be little doubt but that NASA is the public exemplification of this forefront." One outcome of that popular conception, they said, was the idea that NASA presented "a new era of equality according to ability"[13] and that "communities with advanced types of industry, with their people employed in research laboratories and in the development of new engineering techniques, should display a high level of social innovation."[14] They wanted to understand whether those assumptions held any truth, so they sent a researcher out to the space communities to find out. The dispatch that he sent back is fascinating.

The academy's investigator was a sociologist named Peter Dodd. He traveled to the South during the depths of the civil rights battle and paid multiple visits to Brevard County, Florida, the home of Cape Canaveral; Huntsville, Alabama; Hancock and Pearl River Counties, Mississippi, home of the Mississippi Test Facility (MTF); and the Manned Spacecraft Center (MSC) in Houston. There he talked to "technologists and technicians employed by the space centers, administrative officials at the space centers, municipal officials, city planners, newspapermen, ministers, educators, social workers, housewives and teenagers."[15] Dodd spent most of his childhood in New England. He entered Princeton University when he was fifteen and then went on to graduate school at Harvard. Nothing in that upbringing prepared him for what he found during his travels in the Deep South. Dodd went on to spend most of his career in the Middle East as a cultural ambassador, trying to get Muslim cultures and the West to understand each other better. His report from the South in the early 1960s showed he could have used a cultural ambassador himself.[16]

It is evident that Dodd believed the pre-impression about NASA — that people employed doing research into the development of new engineering techniques would display a high level of social innovation. It is also apparent that he believed, as George Carruthers did, that "most of the people who are doing science and engineering are less likely to do segregation than people out on the street."[17] Dodd expected to find towns populated by sensible people with "liberal attitudes."[18] This only made sense, he thought, since "they have high levels of education, which are known to be correlated with liberal views," and "their youth and geographic mobility have exposed them to liberal opinions."[19] To his dismay, what he found was very much the opposite. Instead of a technocratic utopia, Dodd was shocked to discover that "technological advances do not necessarily lead to major changes in social institutions. In fact technological advancement may be associated with a conservatism in social attitudes." In Huntsville, he found the public school curriculum to be "a standard one, without the new courses in mathematics and physics." In a community with many families from Germany, he was saddened to find that "the main high school offers no instruction in German," and that there was no language instruction at all in elementary school.[20] But the middlebrow curricula were only the start of it. He was shocked to find that people in the South were religious, and that they did not practice the liberal branches of Presbyterianism he knew from back home in Connecticut. "In matters of belief," Dodd reported, even engineers and scientists at the space centers "take a conservative position, insisting on a literal interpretation of the Bible. One might expect the incoming technologists to find this traditional attitude to religion difficult to accept. Their education and technological training might lead them to question traditional doctrinal belief." Engineers and technicians, he thought, should be more likely to favor churches that offer "a rational approach to the religious questions, such as the Unitarians." What he found instead was that "the evidence for these communities does not show any 'flight from religion.' The churches, especially in Huntsville, are flourishing."[21]

New England parochialism aside, what astounded Dodd most was the attitude in the space communities toward race. In Huntsville and at Cape Canaveral, he said, "there seems to be no evidence of strong pressure for Negro rights, nor of strong sympathy among technologists for civil rights." To NASA workers, he found, "the Negroes appear to be an outside group presenting demands which would have to be dealt with in some way, but which are no concern of theirs." Blacks in the South, he said, were treated like labor unions in the North, "possibly a rightful movement, possibly a nuisance that would interfere with work, but not possessing any ideological significance."[22] He also found that the NASA workers were the most open-minded of the people he contacted. In the rest of the community, the predominant view was, he found, that "the Negroes should be kept in their place." It is no surprise that what he termed "these 'conservative' views" were expressed "more often and more vehemently in southern Mississippi." In the community outside of the space centers there he found that the view was "the Negroes are being stirred up by outsiders, not excluding Federal government officials." The space center in Mississippi "is seen as a possible enemy of local interests, even though it brings prosperity to the community."[23]

As far as Dodd saw it, these space workers would not be able to overcome "the conflict between their Southern origins and their experience and education at work" and would "not be likely to lead the movement for integration." The impetus for that, he said, "will come from sources outside the community."[24] Years later, there remains a popular misconception that that is precisely what happened; that in the later 1960s, a northern occupation force of scientists and technicians came into the space communities and were responsible for changes in race relations there. This idea not only crops up in oral histories, but in published histories, too. Asked if NASA changed race relations in the South, Julius Montgomery said of Florida, "I guess it did help in some ways, because guys came from the North." In his book *Blacks and Social Change*, James W. Button quoted an unnamed city official from Theodis Ray's hometown of Titusville who claimed that "the driving force behind integration here was the influence and growth of NASA . . . It brought a whole bunch of Yankees to the South."[25] Dodd himself acknowledged that this common misperception existed. He quoted from David S. Akens' 1959 book *Rocket City, USA*, which asserted that "75% of the population [of Huntsville] represents 'immigrants' from all parts of the country, particularly the northern states."[26]

In doing actual research, however, Dodd found that was not the case. "The space centers do not draw their employees, proportionately, from the different areas of the country," he found. "They recruit heavily from technicians and technologists raised and trained in the South. The personnel chief of the Huntsville Center estimated that 50% of its employees come from residences in Alabama alone. He also provided a list of birthplaces of the center's employees. A rough check of this list showed that not more than one-fifth came from outside an area including the southern and Border States.[27] Dodd also concluded that the population of Brevard County was largely southern in origin. Florida state researchers subsequently confirmed this finding, he said. Their surveys "showed that well over half of the migrants into Brevard County had in fact spent most of their lives in the South."[28] According to economist Mary A. Holman, 37.8 percent of the people working in Madison County (Huntsville) in 1965 worked in either Alabama or Madison County in 1960, 17.6 percent worked elsewhere (including the military), and 44.6 percent were unemployed in 1960. She estimated that 30 percent of the labor force constituted out-of-state in-migration to Huntsville.[29] She also found that only 28.4 percent of people employed in Brevard County in 1965 worked outside of Florida in 1960.[30]

These numbers are also important considering that both custom and law accounted for the enforcement of segregation in the South, and that many NASA and aerospace immigrants from outside had no experience with the social and political traditions of small towns in the Deep South and no interest in participating in them. Newcomers to Brevard County, for example, expressed no particular militancy on social issues and lacked both the time and inclination to assume political control of communities.[31] Consequently, the Titusville City Council, according to Button, was "composed of an old-line group of people—rural, southern, here all their lives, and some of whom still carried Civil War memories."[32] It is another manifestation of the Space Age as deus ex machina that what citizens and civic leaders believed about NASA was often as important as reality. In effect, NASA became the scapegoat of southern traditionalists eager to deny a local cause of change.[33] The mythic legions of rocket scientists-cum-social engineers were a convenient conceit for all involved. Traditionalists blamed "Yankees" and other outsiders, rather than their longtime neighbors, for changing a way of life; reformers used the myth to shame the federal government into action; and most important for NASA, the agency's public image remained intact, though maybe not when it came to the thinkers at the National Academy of Arts and Sciences who had asked NASA to fund Dodd's study.[34]

"The Form Was Lost [and] It Has Never Been Found"

These numbers also matter in analyses of race relations within NASA centers in the South when it came to the imposition of and compliance with federal equal employment and integration policies. First, on integration, Clyde Foster said it was an accepted fact of everyday life at NASA in Alabama, just as it was for Julius Montgomery in Florida, that no one would invite black employees to office social functions like birthday and going-away parties. "It sort of was understood," Foster said. "I never attempted to go to any of them. I just knew it was happening and I just didn't subject myself to whatever might have happened had I gone."[35] There were also examples of this type of behavior with much more consequential impact.

For one, the personnel office at MSFC was notorious among its African American employees. Numerous discrimination hearings confirmed this characterization. A hearing in 1962, for example, involved a man named Edward Earl Morton, a pastor at the Union Hill Cumberland Presbyterian Church who had worked at MSFC for nineteen years, first with the army and then NASA. Morton filed a complaint with the PCEEO in Washington charging that the MSFC placement office discriminated against him because of his race. Morton had applied for a truck driver/messenger job and for a job as a mail clerk. With the mail clerk job, the hearing report said personnel turned him down without interviewing him. After his paperwork was sent back to the proper office at MSFC, "the form was lost [and] it has never been found." A few days later, a white man applied for the same job, personnel interviewed him, and he filled the position. Around the same time, Morton applied for the truck driver/messenger job. He took a civil service exam and received a rating of "eligible." A copy of the rating notice, along with the exam, went to the employment office at MSFC "with a cover memorandum requesting that these papers be placed in Morton's file. No one in the placement section was able to tell us where to find these papers, nor were they able to locate such papers when asked to do so." After it was incorrectly determined that Morton was not qualified for the position, someone removed his name from the eligibility list in the job opportunity file. The hearing officer in the case, MSFC chief counsel W. E. Guilian, said, "The loss or overlooking of any one of these items, standing alone, would not, in my opinion, be adequate to support a discrimination charge"; however, taken together "they raise a very serious question." He also added an ominous note, pointing out that if an investigator for the PCEEO ever came down to MSFC to investigate Morton's complaint, "it is my belief that such investigator could conclude, not unreasonably, that Morton was denied equal opportunity because of his race."[36] This was not the only case of this type.

Around the time of the 1963 March on Washington, NASA held a discrimination hearing over the cases of two MSFC employees, Joseph Ben Curry and Joe D. Haynes. In this case, the hearing officer was neither a NASA employee nor an Alabaman. Haynes had charged "discrimination barring promotion" and Curry "complained of assignments inappropriate to his job classification."[37] The hearing officer, California law professor E. J. (Jack) Spielman, made "a finding of discrimination" and asked that the MSFC promote Haynes and give Curry work "consistent with his experience and qualifications."[38] On his copy of the hearing report, MSFC director Wernher von Braun circled the word "discrimination," drew an arrow to the margin, and wrote "color?" On the letter back to headquarters reporting that MSFC had carried out the promotions, there is a handwritten note at the top that reads, "Dr. von Braun—both Curry and Haynes are Negroes."[39] Several weeks later, von Braun wrote to headquarters about the case. His letter was not to report on how the men were doing in their new positions or about changes MSFC made in response to the charges. Instead, he wrote to complain about Spielman, the hearing officer. While von Braun said he had "no quarrel with the decision" in the case, "it seems to me that some of the statements made by the hearing officer were gratuitous and unwarranted under the circumstances." What appears to have gotten under von Braun's skin in particular was the finding by Spielman that "discrimination continues to exist" at MSFC. The numbers that he cited were correct, von Braun said; there were in fact very few African Americans working at the center, but von Braun told NASA administrator James Webb that the raw numbers were the wrong place to look in determining whether discrimination continued at MSFC.

The hearing officer's findings, von Braun said, "failed to reflect MSFC's attempts to encourage Negroes and other minority groups to seek employment." Spielman also refused to accept NASA's principal explanations for MSFC's lack of African American employees—that there were too few qualified African Americans living in the area and that "those employed elsewhere are reluctant to move here." The other thing that angered von Braun was that Spielman disparaged the testimony of the MSFC's principal government witnesses, saying their views were "not worthy of credibility." That charge was "unnecessary and unwarranted," von Braun said. The witnesses were all "men of the highest integrity." Their testimony reflected the information they had received. "These conclusions may have been wrong," von Braun wrote, "and may have been in conflict with the hearing officer's conclusions, but reasonable men should be able to differ on such subjective matters without one impugning the integrity of the other." Von Braun recognized that calling out the hearing officer in this way was an "extraordinary action," but he said he did it because "I am concerned with the effect of this decision on my attempts to promote equal employment opportunity at the center."[40] This comment is an apparent reflection of the Civil Service Commission survey mentioned earlier showing that African Americans "almost without exception, were suspicious of the EEO policy" at NASA. Imagine how much harder it would be to convince people otherwise if a federal mediator was accusing MSFC personnel of insensitivity to racial conditions if not outright lying.

Von Braun requested that Spielman never chair a discrimination hearing at NASA again. The letter back from headquarters was apologetic to von Braun and vowed that Spielman would not return.[41] This episode demonstrates in microcosm an entire portion of the story of NASA and civil rights, reflecting as it does the question of whether the agency and the entire Kennedy administration did enough and whether it could be seen as ever doing enough to bring about equality between the races. The exchange of letters demonstrates the common NASA tendency in situations like these. In fact, it would be difficult to find a clearer demonstration of "bureaucratic caution and political interference" triumphing over "the normalization of risk." Certainly, NASA was covering up to some extent. Smoothed feelings between headquarters and a center director were more important than doing what was right. The letters also show that where people stood on this question depended largely on where they sat.

"Men of the Highest Integrity"

People at MSFC saw the racial climate in the Huntsville area, in Harry Gorman's words, as one of the most "progressive in the South, if not in the country." But Spielman was not from Huntsville. Neither was Rice University president Kenneth Pitzer, who found the racial wording in the documents that founded his school "just ridiculous." Neither was sociologist Peter Dodd, who was shocked to find "no evidence of strong pressure for Negro rights, nor of strong sympathy among technologists for civil rights" in NASA's southern host communities. In the case of Spielman, he looked at Alabama, conflated his impressions with those about Huntsville, and simply could not fathom the place he was visiting. Those inside MSFC, on the other hand, saw the officials who testified as "men of the highest integrity" who were telling the truth as they saw it. Spielman saw a bunch of southern white men. Whites inside MSFC saw themselves as taking many steps to develop an affirmative action program for equal employment opportunity. To Curry and Haynes the situation looked quite different. The racial climate in Huntsville would have yet another significant impact on African American employees at NASA. It touched Clyde Foster directly, and he fought for years to try to solve it.

NASA, as a federal entity, could not discriminate. But even with all the protections in place, NASA in the South was still in the South. In 1962, African American NASA employees could do math and build circuits next to white people, but once they walked off the base, the reality was that Delano Hyder or George Carruthers or Julius Montgomery could not use the same toilet as their coworkers. More importantly to Clyde Foster, they could not go to the same hotels. If a black NASA employee wanted to move from the GS-3 salary range to maybe GS-8 or GS-12—to become a supervisor or get a more challenging or interesting job—"there were some prerequisites," Foster said, specifically, "more training and more education." NASA had an advanced training program. But, Foster said, "One of the things that bugged me the most" was "when you had to go off-post for a seminar or for training, because you weren't allowed in public facilities." Where did NASA hold its advanced training classes and seminars in Alabama? In public facilities, specifically, the ballrooms of Huntsville's whites-only hotels. As if that was not galling enough, NASA gave African American employees time off to take training classes in states that did allow blacks in public facilities. Foster went once to telemetry training in Atlanta. The only place he was allowed to stay was "one of those fly-by-night hotels" that was several miles away from the training center, "but every morning I got up and I didn't miss a class."[42] Foster had a young family then and did not have the time to keep running back and forth to Georgia to take classes in order to advance. So for a long time, he just swallowed his anger and pressed on.

"A Piece of Dynamite Ready to Explode"

But with Martin Luther King Jr. campaigning for so many changes in Alabama at the time—reminding Foster of what might be—his frustration was taking a toll. Living every day knowing people were keeping him down because of the color of his skin, Foster said, "You feel like you are a piece of dynamite ready to explode. The only thing it takes is but a little fire—a cigarette butt—and light the fuse." His explosion came one day at work. "I was pretty good in the computation laboratory; and I had one of my supervisors tell me he wanted me to train another person in that unit [a white man] and 'I want to make him your boss.' That bomb went off—my bomb. He didn't get any training from me and never did he become my damn boss!" Instead, Foster went over his supervisor's head for the benefit of himself and the other black employees at MSFC. "I went to the officials at NASA and informed them," he said, that blacks wanted training held in places they could attend. He and the head of the MSFC worked out a solution, one that seems astounding to us but that made sense in that time and place. They created another training program—separate but equal—held at Alabama A&M and attended by the black employees. "They would import some instructors as far away as Nashville, Tennessee, and we would tell them what kind of courses we wanted to get some training in. NASA paid those individuals to come down and do that."[43] The arrangement perpetuated segregation, but that concern was secondary to Foster. What he cared about was that men he worked with, friends like Hall, Hyder—and all the others who did not have grades like David Mc-Glathery, which would allow them admission to UA-HC—could advance, too, and they did. Thanks to Foster, close to one hundred black MSFC employees took advanced training and qualified for better jobs.

In addition to this, the agency made other strides to atone for its past discrimination and change the nature of its southern workforce and its home communities. The greatest number of changes began shortly after a May 29, 1963, PCEEO meeting in Washington, where

Attorney General Robert Kennedy (RFK) exploded after learning that in the entire city of Birmingham only fifteen African Americans worked for the federal government.

The rioting in Birmingham deeply frightened the Kennedy brothers. It was not the sight of police officers attacking with dogs and rolling children down the street with fire hoses. That was merely disgusting. What had them petrified were the stories of what happened when Birmingham's African Americans started fighting back—what the *Chicago Defender* termed "racial rioting," where "hundreds of enraged Negroes fought battles with police, using their bare knuckles, rocks, bricks, clubs and in several cases knives." In the mayhem, buildings were "gutted by fire," a policeman was stabbed in the back, and a white cab driver was pulled out of his car and beaten."[44] RFK foresaw in Birmingham the burning of America's cities that would come over the next few years. Trying to hold off the rioting during his brother's term, he called a meeting of black intellectuals to get some advice and maybe spread some calming words to ward off the fire next time. Instead, at the meeting, Loraine Hansberry, author of *A Raisin in the Sun*, told him, "We [black America] are without any hope at all. You and your brother are representatives of the best that a white America can offer, and if you are insensitive to this, then there is no alternative except going in the streets."[45]

RFK and his brother saw themselves as having gone out on a limb for the African American community. They risked their political standing—had, in fact cut ties—with a significant wing of their political coalition. But clearly, the very people they thought they were helping gave them no credit. RFK was determined to make sure that the rest of the federal establishment knew what he now knew and that they were doing something about it. At the May 29 PCEEO meeting he came to better understand the root of the problem. The PCEEO and the administration's Plans for Progress initiative with private industry were the president's tools to force integration. What RFK learned at the May 29 meeting stunned him, and as he put the facts together, it started to make sense why businessmen in Birmingham were constantly confronting him, harping on the federal government's lamentable record on hiring. "Why should we hire Negroes?" he said they asked him. "You don't hire Negroes."[46] Vice President Johnson, the chairman of the PCEEO, had been lying to him all this time about the PCEEO's progress, Kennedy thought, while he and his brother were taking the heat.

"Several of Our Key Engineering Personnel Are Colored"

The PCEEO dispute was not the start of the bad blood between RFK and LBJ. That had started years ago, but the dispute did give the feud a specific locus and focus. RFK's anti-Johnson allies included the PCEEO staff director John G. Feild, who hated Johnson's instinct for conciliation and often went behind his back, and labor secretary Willard Wirtz, who liked planting doubt in the attorney general by telling him, for example, that "two-thirds of the companies holding government contracts still [do] not employ blacks." The attorney general passed these assessments along to his brother, who called Johnson "deplorable," adding, "That man can't run this Committee."[47]

Johnson was not at the May 29 meeting, but he heard about it and called in NASA administrator James Webb afterward for a talk.[48] NASA and its contractors had thousands of jobs in the South, particularly in Alabama. With an urgent need to show some progress, Johnson told Webb he needed to know what NASA was doing on hiring. There was another PCEEO meeting coming up on July 18. The attorney general would no doubt show up again, Johnson told Webb. He would want answers and he would want action. Johnson needed to be ready. The meeting started a flurry at NASA. First, at the agency's senior staff meeting, on June 5, Webb issued a startling, desperate, and fairly transparent order. Each member of senior management, he said, had to pursue the goal of recruiting "at least one nonwhite person to their office during the next six months."[49] Webb sent out a letter the next day to every "important NASA contractor" telling them about a compliance survey the PCEEO would be conducting and asking for "your assurance of complete observance of the spirit and intent of the equal employment program."[50]

Webb learned about the compliance surveys from Vice President Johnson in another meeting, held just before RFK tore into Hobart Taylor Jr. Johnson told him that because the question of civil rights enforcement in federal contracts "is constantly emerging as an issue," NASA might be called before congressional committees "and personally questioned as to what we were doing to enforce the law, Executive

Order, and the Plan [for Progress]." The vice president also told Webb about a significant number of backlogged EEO complaints, four hundred of which were against the Department of Defense. Knowing that NASA and the Pentagon shared a large number of contractors, Webb delegated the issue to NASA assistant administrator Robert Seamans, saying, "I think we should find out how many [backlogged complaints] there are in NASA and put on to this effort enough manpower to see that we do everything that is required."[51] He also suggested that contractors provide written statements of their equal employment actions, confident that such a record would be useful if he ever had to testify.

There is no evidence that Seamans had actually gathered any of those statements a few days later when he made a confident and very inaccurate report to Congress on NASA's progress in hiring African Americans at the Marshall Space Flight Center. In testimony before the Committee on Aeronautical and Space Sciences, which was considering NASA's 1964 authorization, Seamans had the following exchange with the chairman, Senator Clinton P. Anderson of New Mexico:

THE CHAIRMAN: You mean they [African Americans] are not just given the janitor jobs? They work as mechanics if they are trained in that field?

DR. SEAMANS: Yes, and secretaries and technical personnel. I can't give you the exact number but several of our key engineering personnel at Huntsville, Ala., are colored.[52]

Neither Anderson nor any other committee member pressed Seamans for a more exact accounting of African American personnel in Huntsville, which was a good thing for Seamans. There were no black "key engineering personnel" in Huntsville in June 1963. There were no secretaries, either. There were blacks doing menial work for NASA in Huntsville at the time of Seamans' testimony. There were also people like Clyde Foster and David McGlathery, who were technicians, but not professionals. The fact is that at the time Seamans testified, MSFC had no black professionals at all.[53] Seamans was the man Webb relied on for this kind of information, too, and this false assessment would come back to bite him soon enough.

Johnson's arm-twisting compelled the agency to take additional small steps to comply with the evolving ethos on civil rights by, for example, tightening its rules regarding the places employees could stay while on the road. NASA changed the wording of a management manual circular from this period to require the following:

> Installation offices placing reservations at hotels, motels, restaurants and recreational facilities shall limit such reservations to those places of public accommodation where NASA employees, other Government employees, and guests of NASA will be served without discrimination as to race, color, creed, or national origin.[54]

In June 1963, it also became against agency rules to speak to organizations that discriminated. Back in 1961 President Kennedy had issued the "Memorandum on Racial or Other Discrimination in Federal Employee Recreational Associations" to make sure that "[there was] no use of the name, sponsorship, facilities, or activity of any Executive Department or Agency by or for any employee recreational organization practicing discrimination based on race, creed, color, or national origin."[55] In 1963, he extended the restriction to public-speaking engagements. NASA quickly incorporated the speaking guidelines into its management manual and created a form letter that allowed it to turn down invitations quickly.[56]

"The Vice President Has Expressed Considerable Concern"

By the end of the month, Johnson had finished his work and NASA's work would begin as Webb began twisting the arm of Wernher von Braun. On June 24, Webb sent a letter where the pressure is evident in the first sentence. "The Vice President has expressed considerable concern over the lack of equal employment opportunity for Negroes in Huntsville, Alabama," Webb said, and noted that he had called a meeting "to formulate affirmative plans for correcting the situation." In that meeting, Johnson came together with Webb, PCEEO representatives from the Pentagon, and the Civil Service Commission, the letter said. They came up with a number of assignments that von Braun now had to carry out.

First, there was an audit coming. "The Civil Service Commission was instructed to survey the federal employment practices in the

Huntsville area at the earliest possible date." Second, because NASA had been unable to find African Americans to fill technology positions, it was now going to have to grow its own. Webb told von Braun to take the lead in helping Alabama A&M College and Tuskegee Institute write grant proposals that might lead to "the improvement of curriculum and faculty to provide better-qualified graduates for employment with NASA at Huntsville." The PCEEO had already been in touch with the colleges' presidents, von Braun was told. The schools were told that "NASA representatives will be calling on them to assist in the development of grants in aid or contract proposals." Webb told von Braun, "Your office should take the initiative in the development of these proposals."[57]

Next, the letter told von Braun that NASA was to call a meeting of all its Huntsville contractors. There, the twisting of arms would continue, as NASA would gather from the contractors "first-hand reports on what they plan to do to assure equal employment opportunity to Negroes in their companies."[58] This meeting would not be held in Alabama but at the U.S. Department of Labor in Washington, DC. Three of von Braun's deputies were ordered to Washington for the meeting, and the letter told von Braun to "personally meet" with the Huntsville contractors "in your office" ahead of the DC meeting in order to review "their plans for assuring equal employment opportunity for Negroes in their organizations." Additionally, the vice president had ordered NASA, along with the Pentagon, to "conduct a housing survey of the Huntsville area" and to submit a report on how the government might ensure that there would be "adequate housing for professional level Negro employees in the Huntsville area." Finally, Webb told von Braun to examine his center's "summer employment situation and take necessary action to see the qualified Negroes are offered a fair proportion of the summer jobs."[59]

The administration had, to say the least, an odd partner for these actions in the person of Wernher von Braun. The word "ironic" does not begin to capture what it meant that a man tasked with implementing a program of racial equality had once worked for Hitler. Von Braun had used the slave labor of concentration camp inmates to build the V2 rockets that fell on London and elsewhere. Yet this was the man tasked with correcting the legacy of slavery in Alabama. In a 2008 biography, Michael Neufeld said there is no indication the issue of race relations bothered von Braun "and ample evidence that Webb was pressuring him, often in a fairly condescending way."[60] There is a more hagiographic portrayal of von Braun that suggests otherwise, but it contains several second- and third-hand anecdotes, and its quotes of von Braun saying things like "I'm not going to sit quiet on a major issue like segregation" are vaguely sourced.[61]

Association of Huntsville Area Contractors (AHAC)

Whether or not it was von Braun's personal doing, NASA took significant, tangible action in Huntsville following Webb's letter. The first achievement came eleven days after the letter arrived, when the Huntsville contractors gathered to prepare for their impending meeting in Washington. The group created the Huntsville Contractors Equal Employment Opportunity Committee and appointed Milton K. Cummings of Brown Engineering as its spokesman. It also created an organizing committee with representatives from IBM, SPACO, and Hayes International Corp. The goal of the organization was to "improve the percentage of minority employment in the Huntsville complex."[62] The group came together at the urging of the PCEEO's Hobart Taylor Jr., who believed that contractors could influence local builders and financial institutions to effect better housing. He also believed that local representatives were better suited to deal with the community at large than were persons from an agency or office without community ties. Taylor suggested using the stick of large depositors and the carrot of economic growth to influence financial institutions.[63] In short order, the group changed its name to the Association of Huntsville Area Contractors (AHAC).

The group's founding document is a demonstration of just how far Alabama was from any reasonable expectation that it could find something close to numerical parity in the workplace between blacks and whites. The executives acknowledged that their companies "received very few applications for Negroes in the professional and technical field." While they said that this reflected "a weakness in Negro educational opportunities," they also acknowledged that "many Negro engineers would not come to Huntsville." It is telling that the group did not propose to go out and hire African Americans from the surrounding area. Instead, they first pledged that they would "aggressively seek out Negro . . . people throughout the United States." And even there, the point was not direct hiring but rather establishing a "working relationship with Negro educational and training institutions and work with their placement officials." The idea was that members would go to black colleges and "point out the areas of greatest employment opportunities" to the administrators at those schools and then "take steps to assist them in training their students in these courses."[64]

The executives also agreed to hire African Americans in their own personnel departments. They declared their understanding that doing this would ensure full compliance with the presidents's executive order and make sure that "testing and qualification standards are administered fairly." In doing so, they demonstrated that they understood how pervasive racism and discrimination were in their communities, a very enlightened position to take in Alabama in 1963. They also vowed to create apprenticeship programs, a desperate need, as labor unions were notoriously discriminatory and apprenticeship positions were practically impossible for African Americans to obtain. In what can be seen as perhaps the most bold challenge made to George Wallace's idea of keeping blacks in a "separate racial station," the contractors agreed to use their influence "in the community and in the state" to make people "more conscious of our responsibility in the area of housing, education and the availability of private and public facilities." In other words, they were going to challenge discrimination and they were going to use their money and their influence to do so. They also said, "We should use our influence . . . to help the colored high schools achieve equality with the white high schools in facilities and instruction." Coming nine years after the Supreme Court's *Brown v. Board of Education* decision, this promise shows just how far Alabama had to come even to comply with the law as written. The contractors also vowed to prepare their employees "for the additional Negroes that will be joining the companies in the future." This is also a breathtaking promise to have made in Alabama in 1963. Blacks and whites still could not use the same water fountains or cafeterias in most workplaces. It is somewhat staggering to contemplate just what "preparing their employees" meant in that context.[65]

In a letter back to Webb on July 5, von Braun offered a progress report on the work of the Huntsville contractors and the Marshall Center. NASA employees, he said, had met with a state senator and the

Alabama superintendent of schools "to point out some apparent inadequacies in the local Negro educational institutions." They recommended, among other things, that fund-raising begin to build a library at A&M and that "qualified white personnel" be allowed to serve on the school's instructional staff. The MSFC's lobbying had apparently been successful, because von Braun reported the college was getting an additional state appropriation of $200,000 for 1964 and an additional $220,000 for 1965. The state had also appropriated $900,000 to build a library. Von Braun's letter also reported on what would come to be a significant program, telling Webb that representatives of twelve "Negro colleges" were coming to Marshall to discuss their interests in creating co-op programs. He said, "It is our plan to have the initial group of Negro co-ops enter on duty this September."[66]

L. C. McMillan, a black college administrator from Prairie View A&M in Texas, was eventually appointed executive secretary of the contractors group. He credited government work in Huntsville as an accelerator for desegregation. Member corporations, familiar with the space agency's emphasis on minority employment, sought to increase the number of black job applicants for technical positions and to facilitate better relations between colleges and contractors. NASA officials at MSFC worked with black colleges on grant proposals and recruitment of students for NASA employment. In a letter to Senator Jacob Javits, Paul G. Dembling, NASA's director of legislative affairs, explained how AHAC worked with federal housing authorities regarding low- and middle-income housing and equal consideration for home loans by lending institutions. By the end of 1963 two of Huntsville's better hotels, the Sahara and the King's Inn, had desegregated.[67] And progress continued. An ad hoc businessmen's group issued a report in 1964 that listed Huntsville as "partially desegregated" in the areas of restaurants, hotels-motels, theaters, public schools, and municipal transportation-recreation. Lunch counters were marked as substantially desegregated. None of Alabama's other major cities had a comparable record.[68]

"I Don't See That the Job Will Be Done"

The actions taken by NASA in the six weeks after the May 29 PCEEO meeting were as impressive to Hobart Taylor Jr. and Lyndon

Johnson as they were to NASA's senior executives. Surely, they thought, at the next meeting RFK would be as impressed as they were. It was not to be. Two weeks after AHAC's creation, the confrontation Johnson feared between RFK and Webb came about in an explosion of power and ego that is perhaps the best-known intersection of NASA and the civil rights battle.

The PCEEO met in a conference room at the State Department. The transcript reads as if nothing unusual was going on. The transcript, however, offers only a taste of the scene in the room that day. The release in the 1990s of Lyndon Johnson's White House phone recordings reveals what Robert Kennedy sounded like when he was angry, particularly when he was angry with Lyndon Johnson. In these conversations Kennedy can sometimes sounds peevish—almost depressed, with a petulant tone that sometimes barely rises above a murmur. But when he is angry, he snaps off his sentences in short clips that drip with haughty scorn, as his Boston Brahmin accent gets thicker. Reading with Kennedy's voice in mind better conveys the tension in the meeting that day. In an interview for the RFK Oral History Project, Jack Conway, a longtime force in the labor movement, declared it "a pretty brutal performance."[69]

The meeting started poorly for Webb. A week earlier, Hobart Taylor Jr. had given Johnson the results of an investigation of employment opportunities for African Americans at MSFC and the Redstone Arsenal in Huntsville. Though 19 percent of Huntsville was African American, the report said, "less than 3 percent of the employees at the above two facilities are Negro."[70] Moreover, Taylor said, "there had been an actual decline in the utilization of non-whites over the past three or four years."[71]

After a brief delineation of these facts, the conversation was shifting to another subject when RFK cut in, demanding details on how Taylor had compiled his numbers in Huntsville. This is when things began to get tense. "Do you have inspectors go around to examine those who have signed up for Plans [for] Progress?" he asked. When Taylor said yes, Kennedy asked to see their reports. Taylor told him to ask the Pentagon's PCEEO representative, Adam Yarmolinsky, because he was the one who heard from the inspectors.

In an attempt to clear things up, Johnson jumped in and gave Kennedy exactly what he did not want—more generalities. LBJ volunteered that he knew of seventy-five companies participating in Plans for Progress that had made a "substantial increase" in their hiring of African Americans. That did not satisfy the attorney general, who began peppering Yarmolinsky with questions on how the Pentagon conducted its compliance surveys. Johnson cut in:

> THE VICE PRESIDENT: How many people roughly do you have working in this field in the Department of Defense?
> MR. YARMOLINSKY: At the present time we have about 50.
> THE VICE PRESIDENT: Do you plan to bring it up to about 80?
> MR. YARMOLINSKY: We do.[72]

Now that he had a number, RFK bore in. Yarmolinsky had eighty inspectors; what about everyone else in the room? The head of the Atomic Energy Commission volunteered that he had eight. "Could I ask, Mr. Webb, how many people do you have working on your program?" Webb had come prepared with all kinds of facts on NASA's progress in Alabama, but as it happened, Kennedy did not care about any of that. He wanted an answer to the one question Webb could not answer. Webb had no idea how many inspectors NASA had. He turned to an assistant. "We have one-and-a-half-man-years of full-time effort in the headquarters," the man said, "and we have efforts carried out through both personnel offices and procurement offices in the field."

> THE ATTORNEY GENERAL: What is one and a half years?
> MR. HARTSON: One year of a man full time in the Office of Procurement doing nothing but this at all. The other half man year is me.
> THE ATTORNEY GENERAL: What is the value of your Government contracts? First, how many Government—
> MR. HARTSON: It is about 90% of $3.47 billion.
> THE ATTORNEY GENERAL: And you can do that with one and a half men per year?[73]

At this point, Jack Conway said, "Everybody was sweating under the armpits."[74] Webb jumped in to say that the program was not, in fact, run by those one-and-a-half men at headquarters but that NASA had nine centers and "our contract administration is decentralized to those Centers." When Kennedy asked how many people he had work-

ing on it, Webb answered, "I am not sure."[75] Webb tried to change the
subject back to the agency's great success in Huntsville, but Kennedy
would have none of it. "How many do you have being surveyed?" he
demanded. "Do you know how many people you have got doing this?"
Webb said it was not about the number of inspectors but about "the
effectiveness of the operation." Kennedy was not buying it. "Mr. Webb,"
Kennedy snapped, "I just raised a question of whether you can do this
job and run a Center and administer its $3.9 billion worth of contracts
and make sure that Negroes and non-whites have jobs." Webb started
talking about the important work von Braun was doing in Huntsville
and RFK cut him off again. "Mr. Webb—does anybody have specific
and particular responsibility for this program?" Webb stammered
about the center's staffs having reporting systems and Kennedy cut
him off again. "Unless we get down into the specifics," he said, "this
Committee and the President of the United States are interested in
this program, I don't see that the job will be done."[76] RFK "humili-
ated Webb," Jack Conway said, and made the vice president "look like
a fraud."[77] After snapping that "I am trying to ask some questions. I
don't think I am able to get the answers, to tell you the truth," the attorney general got up, walked around the table, shook Conway's hand,
and then left. Webb said nothing for the remainder of the meeting,
which lasted another three hours.

Webb was a prolific memo writer, so it is unusual that there is
nothing of his in the documentary record that addresses this meeting or demonstrates that he did anything because of it. Webb most
likely felt confident that the actions taken by the Huntsville contractors and NASA were going to get the job done. RFK decided to focus
on a narrow measurement of competence (and one of his own, devised
on the spot), the number of inspectors. From the transcript, it appears
that no other metric was going to satisfy him. Perhaps Webb truly believed that it was not a problem to have only one-and-one-half people
at headquarters examining $3.9 billion in contracts for their equal em-
ployment compliance because the centers were able to do the job on
their own. Maybe he knew RFK well enough to know that he was just
blowing off steam and that he was in a no-win situation. For whatever
reason, this meeting passes without further comment in NASA's historical record.

That is not the case in the writings of civil rights historians. There, authors consider this meeting pivotal. The numerous books that mention it attribute the attorney general's anger to his confrontation in New York with the African American intellectuals. Most depict the scene with Johnson and Webb—coming just days after the meeting in New York—as the very height of RFK's pique. Some authors suggest that changes in government hiring in Alabama came about because of this confrontation. The problem is that most of the books that mention this meeting get it wrong. Some have RFK coming straight from the meeting with the black intellectuals and telling Webb on May 29 that he would have him fired—a meeting Webb did not attend. Others conflate the May 29 meeting and the July 18 meeting, saying Kennedy yelled at both Webb and Taylor *together* after meeting with the intellectuals.[78] Considering how much occurred on the civil rights front in 1963, these errors make an enormous difference, as they suggest a cause and effect that did not exist. The Birmingham unrest ended around May 13. Wallace's stand in the schoolhouse door was June 11. The president's civil rights address was June 12, and he introduced his civil rights bill on June 19. Webb engaged in his flurry of civil rights–related correspondence on June 5 and 6, and he sent his letter to von Braun about "the lack of equal employment opportunity for Negroes in Huntsville, Alabama" on June 24. Whether Robert Kennedy threatened to fire James Webb on May 29, June 18, or July 18 makes all the difference in the world.

So many years after the fact, NASA's efforts to enact federal employment policy and establish its own racially progressive programs seem ineffective at best and token at worst. Agency efforts were genuine but incidental to NASA priorities. Indeed, most federal equal employment programs were in their infancy during the 1960s and had limited range and success. It is nonetheless worth looking at what NASA tried to do and at the lives of the people whom this policy touched. The agency and its contractors articulated a commitment at this point in 1963. Now came the hard part, where they would try, and fail, then try again and fail less badly. Little did they know that they were facing forces greater and deeper than they understood—forces scholars would debate for generations.[79] Lyndon Johnson and Robert Kennedy believed (or at least hoped) that the best way to mend the tear in the American fabric caused by the nation's race problem was to open more jobs to African Americans. On the face of it, that is a simple and straightforward matter. Time—and the experiences of NASA's African American pioneers—show that it was, in fact, not straightforward at all.

"Maybe This Is One Way We Can Get More Employees"

As pressure on NASA mounted, Webb ordered von Braun to develop his own cadre of African Americans to fill technology positions. Clyde Foster had no idea about the high-level goings-on, but, as it turned out, he thought NASA growing its own cadre of African American technologists was a good idea, too. In 1964, he took the first steps to see that would happen. Foster was still toiling in the ranks of NASA's nonprofessional employees, but through his work in the computation lab, he said, "I was becoming very much computer literate," to the point where he had made a name for himself in Huntsville's black community. Sometime during the year, an instructor at Alabama A&M contacted him because he "had received a computer but he didn't have anyone to install it. He asked me if I would give him some help." In the course of doing that work, Foster got an idea: he would use his skills in computer programming to train young people in this new, emerging field so that they could get jobs at NASA. "I said, 'Maybe this is one way we can get more employees.'" He took the idea to his supervisor, who took it to von Braun. Here was an opportunity to work directly on "the improvement of curriculum and faculty to provide better qualified graduates for employment with NASA at Huntsville," just as the agency had promised to do. Foster's idea was a natural. As a result, Foster said, NASA "cut a deal that they would place me on loan to the university to establish a BS degree in computer sciences." This was, he said, "the first time anybody had been placed on loan from NASA to the university" and it was "the very first baccalaureate granting program [in computer science] in the state of Alabama."[80] As for the aim of growing a new crop of NASA employees, that worked out, too.

In 1964, James Jennings was majoring in math at A&M when Foster arrived to set up the computing program. "We had the opportunity to take a couple of computer science courses and that's how I kind of decided to get interested in NASA," Jennings said. It was Foster's program at A&M that "sort of kick started my career." He began as a NASA co-op student and then moved into a full-time job writing computer programs to design the trajectory to the moon.[81] At that point, he said,

Foster "kind of took me under his arm and mentored me through the process." As a result, "I didn't have to go through a lot of the things that he did, because he had been there before and could guide me in the right direction." Foster believed in the principle of Each One Teach One, and NASA was the beneficiary. "I often refer to Clyde as my professional father," James Jennings said.[82] That is a strong statement, but it was exactly what Foster had in mind.

Jennings was not the only beneficiary. "The program at Alabama A&M has helped many students get into the technical field," Foster said. "If they didn't get with NASA, they got with the army, and also some of them went into business."[83] This was yet another example of Johnson's plan coming to fruition. "Initially I didn't get the magnitude of what I was doing," Jennings said. "It wasn't until you saw that first launch that you realized that I was a small part of such an enormous part of history."[84]

"Holy Mackerel"

At the same time Clyde Foster was minting new computer programmers for NASA, he was also involved in another, unusual, sidebar activity. In its early days, NASA was very keen to talk about spin-offs from space exploration—added benefits to society that would hitchhike on its rocket ships. Clyde Foster's story is a spin-off from NASA's experience with civil rights. There was an enclave just outside the gates of the Marshall Space Flight Center. Morgan Watson, one of NASA's first black engineers, described it as "an all-black town with outdoor toilets and the whole thing. Civilization hadn't quite made it there yet. That was the Old South."[85] The name of the place was Triana. Clyde Foster lived there and he itched to make it change.

An article in *Ebony* at the time gave a comparable sense of Triana, saying it "belongs more to the antebellum South than the era of astronauts. It is far from Space Aged—a collection of ramshackle farmhouses scattered randomly around two churches and a restaurant. City Hall is a renovated shack heated by a coal stove."[86] There is a picture in the article of that old coal stove. People are holding a town council meeting around it; Clyde Foster chairs that meeting. The article did not say it, but, at the time, that old shack was about to come down. When the article came out, Foster had just secured a grant "to build a brand-new town hall. On one side of the town hall, we had the administration such as the Council offices, fire department, jailhouse, recreation department, and then on the other side we had a day care and gymnasium, recreation for the kids. We also got paved streets, water, lights." The transition from the old town hall to the new mirrored the transition of the old Triana to what it became because of Clyde and Dorothy Foster. From a place with "streets rotted away and chickens peck[ing] sleepily at its only landmark; a pre–Civil War cemetery long covered with leaves,"[87] Triana—through the passion and tenacity of its African American citizens—was transformed into a bedroom community for a new black working and middle class. It also became a template for obtaining a brand new style of African American political empowerment.

The Fosters moved to Triana when they returned to Huntsville from Clyde's stint teaching in Selma. Despite the community's precivilized conditions, Dorothy Foster was happy to be back among family and back among the stories that she knew. One of those stories was about Triana itself—and it was one that Clyde just did not believe. "She was telling me that the area was [once] a city. And I said, 'You're out of your mind, girl, this isn't a city. I don't see any remnants of anything even favoring a city.' Not a stoplight, not a stop sign. And then I started talking with her father, and her father started talking about the time they had police officers, sheriffs, deputy sheriffs who were black."[88]

Though Clyde did not believe it, the story his father-in-law and Dorothy told him was true. Triana had once been a town, and an important one. Not too long after the Revolutionary War, in fact, it was a notorious slave-trading center. Dorothy Foster said, "Wealthy whites bought the land in Triana and brought their slaves with them. The plantations were maintained outside of Triana."[89] Because Triana was near a river, there were many who believed it would "someday overtake Huntsville as northern Alabama's largest city." But the state could never rouse itself to spend money to enlarge the Huntsville Canal, and by the 1820s "the town [had begun] to die." Dorothy Foster said that "when the railroad came in, that's what killed the business in Triana." Whites moved out, she said, and "blacks had no way of keeping all that up and it all deteriorated." As it did, the people there seemed to forget Triana's past. Clyde Foster's skepticism, though, was finally broken down by his in-laws' persistence. They "were so strong on it, to the extent that I said, 'Well I'm going to look into this.' And so, I told the girl to put on her hat and shoes; we're going down to Montgomery, Alabama," home of the state archives.

Pouring through property records from the nineteenth century, Foster found something that astounded him. There had in fact once been a city called Triana, Alabama, officially chartered in 1819. "And I said, 'Holy mackerel.'"[90] Although the town of Triana had died, no one had bothered to dissolve its charter. Looking further, he discovered a loophole in the Alabama state code that he thought might allow him to improve conditions in his town.[91] There were many reasons why Huntsville's boom had bypassed Triana. Of course, a big part was because it was 86 percent African American, but there was also the issue of leverage.[92] Alabama neglected Triana because it had no influence. It had no influence because it did not legally exist, or at least people did not think it did. What the Fosters discovered in the Alabama State Archives changed all that. For the people of Triana, it changed everything.

Before the trip to Montgomery, Foster had been working to get Triana running water. He tried to start a water authority and get the U.S. Department of Agriculture to finance it through loans. Now, though, "when I found out this was a town, I said the easiest way to do this is: let's incorporate the town, and the town will now be eligible for some grants that you don't have to pay back. For the same price."[93] With his goal in mind, Foster set about his task.

Reinstating the town was going to take work. A judge had to be convinced and that meant hiring lawyers, gathering signatures, organizing. Foster returned from Montgomery and he and Dorothy spread the word about what they had found, began raising money ("We sold some fish plates, chicken plates, raised enough money"), drew up a petition, and got enough registered voters to sign it that they could go before a judge. Their drive faced stiff opposition. African American self-government was unheard of in 1960s Alabama, and Huntsville's white power structure "bogged the project down in red tape."[94] The white lawyer they had hired as an advisor quit midway, citing "public opinion."[95] Foster, working on his own, persevered for months until finally the judge gave in. He declared that Triana was once again a town, appointed Clyde Foster mayor, and named a five-member town council—all black.

There was not universal celebration of this victory in the black community. "The word was out," Foster said, that "we were [saying] that we wanted to be separate just as King was marching on integration." There were many who said that, under the circumstances, black power just was not necessary. Foster disagreed. "Politics and money goes hand in hand," he said, and he was intent on using these new-found political rights to raise the money that would give the little community the services it needed.[96] It worked. Triana got its U.S. Department of Agriculture grant and built its water system. The town joined the twentieth century sixty-five years late. The lives of Trianans were now on par with those of other people in Huntsville. It does not seem like a lot to expect for Americans to have running water and reliable electricity, like their neighbors, but in the rural South that was not always the case. It took power to make it happen. Triana had never had that before. Thanks to Clyde Foster, it did. Once that became clear, not only did opposition in the black community disappear, but the fight for Triana started a movement.

Hamilton Bims, a longtime reporter and editor at *Ebony*, heard about Foster's victory and brought it to the wider world. He wrote a long article that came out in the magazine's March 1965 issue, and Foster started hearing from other black communities almost immediately. "Oh man," he said, "this set off a bomb!"[97] The decision to grant Triana its sovereignty came at a critical time in the civil rights movement. The president had signed the Civil Rights Act, Congress was debating the Voting Rights Act, and African Americans were hungry to exercise their newly won power. News of Foster's victory gave them the tool they had been looking for as well as a demonstration of why the rights were important in the first place.

It began just in Alabama. "After we got started," Foster said, "a lot of [people] said, 'Hey, we need to go through our history books to see if there are any places in Alabama where there might have been some towns.'" Shortly, he said, "many started developing their own towns." Others followed his example of obtaining political power. "A boy down in Tuskegee—Johnny Ford—he wanted a piece of the pie," Foster said. "And he got to be mayor of Tuskegee." After the movement took hold in Alabama, there was no stopping. "Later on, boys over in Mississippi got in touch with us and they had quite a few black mayors over there [in] towns that were started in the 1800s."

With the gathering steam, Foster realized these new elected officials needed to organize—gain strength in numbers and lobby in their state capitals and Washington. They originally called their organization "'Alabama Elected Officials'—because down south, in that Selma area there had been one or two sheriffs elected" in addition to mayors. After the movement took hold in Mississippi, Foster said, "We started calling ourselves the Eastern Conference of Black Mayors. It started spreading all over. I guess right now there are many more. I sometimes plan on contacting the National Conference of Black Mayors and get an update on who are they and from what area are they."

In order to answer any question about whether NASA had an impact on the way of life in the South, one needs to examine every aspect of life. The agency's role in desegregating workplaces and housing was probably smaller than proponents would have wished. In analyzing NASA's impact on the South, however, how much of a counterweight is the story of Triana? Revived towns with black mayors, black town councils, black sheriffs, and administrators all came about because of a NASA employee—a NASA employee who gave the agency full credit. It was NASA, Foster said, that gave him the "experience earlier in life to cross over to those areas—political, education, business." All of it, he says, "was done because of the experience I had with NASA."[98]

The *Huntsville Times* ran a picture of the town's celebration on July 24, 1969, four days after Apollo 11 returned from the Moon. In it, a group of citizens is hoisting Wernher von Braun on their shoulders downtown. Just in front, by von Braun's right knee, you can see the broad back of Clyde Foster, smiling and clapping at the scene. His own copy of the paper from that day has a big circle around him, with the word "Clyde!" in blue pencil—his own tangible connection to immortality. "The striking thing" that day, Foster said, was that "around the county courthouse at that time there were bales of cotton. Bales of cotton! And here we'd placed a man on the Moon."[99]

Clyde Foster's story demonstrates the interplay of factors that worked together in order to enable NASA to have an impact on the South. Politicians and managers labored to bring change from the top down, while in the communities, passionate African Americans worked assiduously to make life better on their own. All the while, the black press shouted the news. In Foster's battle, a NASA pioneer changed perceptions of what African Americans could accomplish in politics. NASA was a technical organization, the nation's technological elite, and in 1964, the interplay of bottom-up and top-down factors would usher in a remarkable group of young men who would break down an additional wall and change America's perception of the technological prowess of black people in the South.

Water Walkers

Morgan Watson and
George Bourda

We felt that the whole image of black people was riding on us as professionals and we could not fail.

Morgan Watson

When arm-twisting by President Kennedy over equal employment opportunity filtered down through Johnson to NASA administrator James Webb and Wernher von Braun, it came to rest firmly on the shoulders of seven young engineers from Baton Rouge, Louisiana. These men—Walter Applewhite, Wesley Carter, George Bourda, Tommy Dubone, William Winfield, Frank Williams, and Morgan Watson—became the embodiment of Johnson's plan for jobs in the South. They had a significant impact on race relations, coming along as they did at precisely the right time to capitalize on NASA's status and bend it toward the errand of furthering racial equality. "Remember, we are talking about the mid-'60s," Morgan Watson said, "when NASA is making this big thrust forward to get a man on the Moon and on the other hand on the social front, they are making big strides in hiring more black professionals."[1]

"We were glad that God brought us to the time when things were changing," Watson said. That change was all around and he certainly took full advantage of it. Watson grew up in St. Joseph, Louisiana, where, he said, "cotton is king and during my elementary and high school years I picked cotton." Black schools worked on a split schedule back then, he said. "We went during the summers to chop the cotton and during the fall, to pick the cotton. As a result, we had no football teams or a band."[2] Echoing Sonnie Hereford, Watson said that when he was out in the fields, the buses would often roll by filled with white students on the way to classes and activities that southern racism would not allow him to attend. Decades later, when he talks to school groups about those times, the way he puts it is, "Some of the things they'd shout were hurtful and not very nice."[3] It leaves a lot to the imagination, though maybe not the imaginations of the twenty-first century children to whom he is speaking. "Stupid"? A child could imagine that. "Jig!" A child might imagine that too, or worse. The rules, restrictions, and expectations that constituted the southern way of life weighed heavily in those days. Even as a child, though, Morgan Watson was determined that those ways were going to change and that he was going to help. Watson was what was called a "race man" at the time—someone who looked around at the way things were for blacks and said, "Hell, no!"

A Greasy Thumb

Like Otis King, Watson began his activism when he was a teen. On his paper route during the election that brought Eisenhower to Washington, he noticed that in his county "there were no black voters. They would break it down by what the votes were and what the race of the voters was, and in my particular parish there were no black voters." He decided that this wickedness would not stand. "When you're young," he said, "you look for battles to fight." He had a friend whose birthday was the same month as his. "We pledged that when we got eighteen we would go down and register to vote." Jim Crow was the law in Louisiana, which meant a black man needed to know his place. He could not just vote. There had to be hoops to jump though. In Watson's case, it was a literacy test. "We aced it," he said, and he and his friend became voters. It was only one tiny parish in Louisiana. It was only two votes, but they had pierced the southern way of life and Watson remembered that moment as pivotal. "Being one of the first black registered voters in my area sort of just gave us a little impetus to go and expand a little farther."[4]

In 1950s Louisiana, as in the rest of the South, job prospects were dim for African Americans. Watson mirrors what other NASA pioneers have said, declaring, "If you stayed in the South and you went to college, the best you could be would be a teacher." As a result, "all of the people that I knew before me that went to college came back as teachers," he said. Watson's co-pioneer George Bourda saw the same thing. "When we were coming up, blacks were mostly teachers," he said.[5] The alternative was unbearable poverty, so, Watson said, "our parents and teachers and other leaders in the community all pushed us all to go to college." When he was young, Morgan Watson had what he called "a greasy thumb." He liked to tinker with things, taking them apart and putting them back together. "I worked in a hardware store," he said, "and the white store owner saw my report card one day and saw that I had good grades in math and science—chemistry and so forth and he said, 'You know, you would probably make a good engineer.'" Watson did not know what an engineer was so, he said, "I went to the library and started reading about engineers."[6]

Sitting Out the Sit-Ins

He and Bourda ended up at Southern University in Baton Rouge (SU-BR), and, as with everything else, they arrived on campus at an auspicious time. By 1962 the sit-in movement, popularized in Greensboro, North Carolina, had proven its value and had blossomed. An article that year in the *Chicago Defender* said that because of sit-ins, "countless stores, restaurants and railroad and bus terminals are receiving and accommodating Negro patrons on an integrated basis."[7] Success fed on success, and soon earnest young black people all over the country were following the same script: dress up, go downtown, sit at the white lunch counter, order food, get arrested. Students at SU-BR—thirty-five of them—staged the first sit-in in the state of Louisiana. By 1961, the cases of seventeen of those thirty-five had made it all the way to the U.S. Supreme Court.[8] SU-BR students sat-in again in 1962, and when they did, the university capitulated to white community pressure and threw them off campus. The result was that "they had riots and some of the people were tear gassed," George Bourda remembered. "I had one good friend who was arrested and stayed in jail for a little bit."[9] Two thousand students protested and shut the school down. The president of SU-BR resisted change; the black press called him an Uncle Tom and accused him of turning the university into "a vast plantation" of "handkerchief-heads, good only in the vanishing, unskilled act of picking cotton." All across the country, however, university and high school students were, in the words of the *Chicago De-*

fender editorial board, "rapidly reaching the exact degree of intellectual sensitivity which gives them an awareness of their responsibility to the community" and seeing to it that a change would come.[10]

Though Watson and Bourda were concerned about that change, when it came to the SU-BR sit-ins, they sat out. "People in the political science department got involved," Bourda said, and "pre-law students got involved," but "I didn't participate."[11] Watson was coy about his role. "We all had roles to play in the movement—by demonstrating or going to the mass meetings or whatever we were called upon to do. Some became Freedom Riders; others just did various things to open the doors."[12] What he did was continue to study hard and stay out of trouble. That Watson and Bourda declined to participate while a civil rights–related protest consumed their campus is completely counter to what we think we know about that time and place. Those removed by decades know what they know from historians, teachers, and popular culture. Perception is reality, and that reality often becomes the history written by survivors and academics long after the fact. This is true of the civil rights movement of the 1950s and 1960s. The movement was never as monolithic and coordinated as portrayed in a general history textbook. The decision about whether to become involved in the civil rights fight tore apart many black churches and black communities. In Birmingham, for example, at least sixty black churches supported the movement, but more than four hundred did not. The reasons varied—people had philosophical differences and class differences; people feared violence; some were concerned about repercussions. It was the same on black campuses, including SU-BR.[13] Watson and Bourda would have an impact on the struggle for equal rights and the recognition of African Americans, but it would not come while they were in college.

"The Black Happen to Be Charles Smoot"

As the SU-BR sit-ins wound down, NASA took an important step that inadvertently placed Watson and Bourda on the road to their destiny. It happened in 1963 when the Marshall Space Flight Center (MSFC) hired a man named Charles Smoot. Smoot was teaching high school chemistry when he applied at NASA. He hoped for a job "on the technical side," but to the benefit of many, as it would turn out, the agency found a place for him in personnel. The job he got seems like an odd one at a technical agency, but at the time, it was necessary. "They needed someone in personnel to interpret certain scientific terms for the personnel people," he said. Word would come down that the agency needed a new employee with expertise in thermodynamics, for example, and, Smoot said, "The personnel people didn't know what it meant." How do you place a want ad in the right publications or contact the proper schools without knowing what skills the technical staff needs? NASA originally hired Charlie Smoot to solve that problem.

Smoot had had the assistance of alert teachers as a child and that help continued when he got to high school. In addition to excelling at math and science, he was good at basketball and, as it turned out, "the coach was a chemist by training." Smoot saw the coach as an inspiration—an adult making a life in the field he loved. "I wanted to be like him. He probably saw something in me and nurtured that." Smoot maintained his studies and, like so many other smart African Americans at that time, he graduated to become a teacher. When his new wife got a job offer in Huntsville, Smoot applied at NASA principally because the salaries were better. "I was hired as a GS-5, $5,500 a year," he remembered. Still, "that was quite an increase from teaching."[14] Fortunately, in Smoot's case, NASA for once applied "the normalization of risk, the acceptance of failure, and the anticipation of trouble" to its personnel operations. Someone got the idea that it made sense to send black recruiters to black schools to talk with potential black recruits. Once that decision was made, Charlie Smoot got a promotion and became "the first Negro recruiter in government service."[15]

Smoot found his first recruit on a trip to Arkansas. E. C. Smith was at the white university there in 1963 working on a master's degree in mathematics when NASA came through. Smith remembered the recruiting team as "two whites and one black." He knew all about Alabama, had seen Bull Conner's police dogs attack, and had watched George Wallace stand in the schoolhouse door; Smith also recalled that nothing those two white men told him had any impact on getting him to work for NASA. But then there was the other man. "The black happen to be Charles Smoot," Smith said, and that made all the difference.[16] "I could be up front" when talking with potential recruits, Smoot said. "They felt comfortable with me as opposed to someone else [who was white]."[17] Smoot gave Smith reassurances he felt he could believe, and those reassurances changed his mind. "From his interview with me," Smith says, "I came over here."[18] This was the hallmark of Smoot's tenure as NASA's African American recruiter. The doubts and concerns that E. C. Smith conveyed were common, but Smoot had a way of talking to young people that overcame their concerns. The reason was simple. "I think it was because I was also a black man," he said. It opened up the conversation in ways a white man could never hope to accomplish.

Smoot traveled the country—going up north to Michigan and out west to California—looking to convince the best African American scientists and engineers to come and work for the space program in Florida, Mississippi, Texas, and Alabama. The job of convincing was "very difficult—not easy at all," he said. They would ask about toilets and drinking fountains; could they use the same ones as their co-workers? They would ask, "How difficult would it be for me to find housing?" Repeatedly, he would give assurances and they would take hold.

In Smoot's parlance, the people he found fell into three categories. There were "regular guys," those who were "smarter than the average bear," and then there were the special ones—the ones Smoot called "Water Walkers."[19] They were scientists and engineers who would be at their top of their field no matter what their race, he said.

"We Were Gonna Change the Whole World"

In the late summer, 1963, Smoot got a new assignment. Back in July, NASA had brought representatives of twelve black colleges to MSFC to talk about establishing co-op programs that would bring promising African American engineers and scientists to work for NASA. "It is our plan," MSFC director Wernher von Braun told headquarters, "to have the initial group of Negro co-ops enter on duty this September."[20] It was Smoot's job to make it happen. He turned to the place he considered most promising, making a phone call to Dr. Dickie Thurman, dean of the College of Engineering at SU-BR. "I liked Dickie," Smoot remembered. "You could communicate with him."

The school had just directed Thurman, an architect by training, to create an engineering program. New opportunities were emerging in the South. Industry was creating new jobs that, even in Louisiana,

blacks could now compete for and get. SU-BR had decided to capitalize on this, and it had worked. Recruiters from aerospace, automotive,
and other manufacturing companies looking to fulfill their own equal
employment opportunity promises by hiring newly minted African
American engineers besieged Thurman's program. Martin Marietta,
Chrysler, Lockheed Missiles and Space, and others lined up at SU-BR,
dangling offers before Thurman's students of summer jobs far away
from the turmoil of the South—jobs that promised more money than
they could make working for the government.

Normally this would have posed a challenge for NASA, but in this
case, the agency had an ace in the hole. That ace was Charlie Smoot. As
he had done numerous times before with African American scientists
and engineers reluctant to move to Alabama, Smoot now brought his
charm and empathy to bear on Thurman. "There was a little bar down
the street," Smoot explained, and he and Thurman would often retire
there for long talks. Smoot was an alcoholic, and though he was in re-
covery at the time, he still loved the atmosphere of the tavern. He felt in
his element there. The two of them, kindred spirits, would spend hours
together. They also shared a vision for advancing African Americans in
technical fields. "We all was young and ambitious at that time," Smoot
said. "We were gonna change the whole world."

They changed the face of NASA instead. The enthusiasm and bon-
homie that Dickie Thurman lavished on Charlie Smoot was not out of
the ordinary. Students said he was like that with everyone. "He could
reach students," Smoot said. "Students could identify with him."[21] That
is also how George Bourda remembered him. "He was a really nice
guy. We used to socialize around his office." With the sit-ins winding
down, things on campus had calmed down and students were turning
their attention back to things like finding a job. Dr. Thurman's office
was a good place for that, too, because "he'd post things on the bulletin board," Bourda said.[22] Bourda and Watson were there the day when
Dr. Thurman posted the offer Charlie Smoot had brought to town.
NASA had established a cooperative education program that would
allow SU-BR students to be in school one semester and work the next
semester at the MSFC. By the time Smoot showed up, Watson was on
his way to becoming one of those "Water Walkers" that Smoot and
NASA prized. In fact, SU-BR considered Watson and Bourda, along
with Frank Williams, Tommy Dubone, and two others, its most prom-

ising engineering students. The conviviality of that little tavern down the street guaranteed that NASA would get the benefit of their talent and skills.

"Negro College Youth to Boost First Moongoer into Orbit?"

The seven young men were given aptitude tests ("no [white] students who were brought as co-ops to Marshall had to take that exam," Smoot said), and once the agency was convinced they were eligible, it hired them.[23] They began at MSFC in the spring semester of 1964. Their assignment was treated with the attendant hoopla the black press normally showered on African Americans working in the space program, and then some. "Negro College Youth to Boost First Moongoer into Orbit?" was the headline in the *Chicago Defender*. The headline employed the common journalistic cheat of placing a question mark at the end of an editorial statement of hope, but the first two sentences made clear where the writer's heart was.[24] "They say there's a good chance that a Negro may be the first man on the moon," the article began. "But even if he isn't, there's a good likelihood that Negro collegian scientists will be performing their intricate duties at the launching pad as they wave the moongoer off." The sentiment is so easy to deconstruct. We all glory in the achievements of the space program, the article is saying, and we all recognize it as a signal American accomplishment—one that our boys get to share. "That was the upshot of an announcement this week that several students of Southern University Baton Rouge, Louisiana had begun their studies at Marshall Space Flight Center in Huntsville, Alabama, where they will study along with scientists during the second semester."[25]

When they entered the NASA co-op program in 1964, Watson, Bourda, Dubone, Winfield, and Williams became the first African Americans ever to work as engineers at NASA in the South. They also constituted a fivefold increase in the number of black professionals at Marshall.[26] "I don't even think there were any clerical workers," Morgan Watson remembered. There were "some groundskeepers and janitors," he said, "but as far as professionals, there were none." The black newspapers were not the only ones who noticed this significant moment. The *New York Times* called Frank Williams "a social pioneer" and "a symbol of the desires and frustrations of space agency offi-cials."[27] Put aside the hype and excitement two years earlier when Ed Dwight found himself two stages away from becoming an astronaut. This truly was an opportunity to "break the ice for others to follow us," as Watson put it, and finally demonstrate that African Americans be-longed at the pinnacle of American science and technology. Was this an opportunity, most definitely, and was there pressure? Yes, and it was immense. "We felt that the whole image of black people [was] riding on us as professionals," Watson said, "and we could not fail."[28]

The thing about being a pioneer of course is that your job is to hack out territory for those who will come after. The pioneers are the ones who endure the rough terrain, hang over chasms, and scale mountains for the benefit of future generations. "We felt that it was our job to blaze the trail on the professional level," Watson said. And although "that's what we focused on," the distractions were plentiful. The *New York Times* got it right when they talked about these young men sym-bolizing "the desires and frustrations of the space agency." Both of those things were there from the beginning.

Whatever the Huntsville contractors or NASA may have said about using their influence "in the community and in the state" to make people "more conscious of our responsibility in the area of housing, education and the availability of private and public facilities," this was, after all, still 1960s Alabama.[29] When Watson, Bourda, and the others got off the bus in Huntsville, their welcome was anything but welcom-ing. "There were no apartments or hotels or anything that would allow us to live there," Watson remembered about his first hours in Alabama. "Everything was segregated." Watson said he watched northern blacks try to make it in the South. They never did because they did not know the rules—keep your head down and go along to get along. Northern blacks "would tend to buck the system, whereas those of us that where from here knew how to tolerate the system and survive." For northern-ers, he said, "coming to the South was like going to a battleground. You would just as soon go to Vietnam and die on foreign soil rather than die in America."[30] The SU-BR men knew the rules, but that was only the beginning of the story.

"A Product of Not Just Themselves"

Charlie Smoot went out before they got there to try to find them accommodations, but the community and the state apparently were just not at all prepared. "Dr. Martin Luther King and the others were doing great things all across the state of Alabama—Selma, Birmingham, and other places," Watson said, and while Huntsville was "sort of off the beaten path of the civil rights movement, still, the effects of segregation and all of the other oppressive issues of that time were upon us." With nowhere else to live, he said, "they found us rooms with people in the community—right in the heart of the black community."[31] As Bourda remembered, "Me and Frank—we were roommates. Tommy lived right next door."[32] Morgan was around the block.

The homes they stayed in were those of the highest echelon of African American society in Huntsville. "They were ministers and teachers," Smoot said, and he selected them for a reason. The families in those homes, in addition to providing a bed and meals, were also "kind of chaperoning them," Smoot said. That is how things worked back then. "If there was a kid who was going places, they'd keep an eye on him. Make sure he didn't get some little girl knocked up, or if they got in trouble with police," someone in authority in the black community was there to step in. "These kids were a product of not just themselves but the product of many people that they don't even remember," Smoot said, which was also the way it worked. "Teachers took care of you, the community took care of you," he said. "It was like a village raising a kid."[33]

Whether the co-op students actually remembered every adult who touched their lives in Huntsville, the time they spent in the city's black community was sweet. "It was like being in an extended family," Watson said. "We got very close to the people. They had children themselves. They would cook for us and everything, so it was just like being at home."[34] As Bourda remembered, "They were some nice people in Alabama—the blacks. We didn't get to know the white ones so much."[35]

"There Was This Rope Right Down the Center of the Arena"

Work was one thing. NASA said they had to work there, but away from NASA, Wallace's "separate racial station" ethos prevailed.[36] "I will never forget one of the concerts that we went to," Watson remembered. "Ray Charles came to Huntsville and there was this rope right down the center of the arena where he performed; all the whites were on one side . . . [and] the blacks were on the other side. And I will never forget when Muhammad Ali [then still Cassius Clay] fought for the heavyweight championship, when he beat Sonny Liston. We all stood out in a big field watching closed-circuit TV. And again it was a segregated situation."[37] That was life in Huntsville. "We'd play basketball," Bourda remembered, and "some of the white kids in the neighborhood would play with us. But then their mothers would call them in. 'You come in now. I don't want you out there with them.'"[38] And that was how things remained.

As Watson remembered, however, "even if [NASA] had the segregated, invitation-only parties and didn't invite us, it didn't make any difference. The social events were kept hush-hush, by invitation only. But there was so much going on in the black community." Besides, he said, "the type of music, the type of entertainment" in the black community "was more down our line anyway." Who needed a bunch of stuffy, white-people's parties? "When we got ready to socialize," Watson said, "we had a lot of social outlets." Tommy had an old Ford Fairlane. Frank had a Buick and then got a new Pontiac. On the weekends, Morgan said, "we would jump in the car and go off to Nashville or over to Atlanta."[39] Plus there was plenty to do right there in Huntsville.

Bourda remembered, "Alabama was kind of different from Louisiana; Louisiana is a real party place." He insisted that "most of the guys—they were not the party types," and on weekends they would mostly "go to football games and stuff like that." Still, they were young, they were smart, they had jobs at NASA, and their names were in the papers. These young men could walk into Adella's Supper Club or Biggun Seay's and they could turn some heads. "We didn't mess with liquor too much," Bourda said, but "we had a good time."[40]

All that attention could give one a swelled head. It could also offer a sense of self that a racist white community might find uppity. "I guess being in Huntsville, we felt kind of bold," Watson said.[41] But in a time when "outside agitators" were disappearing and showing up in ditches or hanging from tree limbs, being "bold" could present a problem. There were times when your life might depend on pretending to "know your place" around the wrong people. As Theodis Ray admonished: "It doesn't sound logical, but it's the truth; each state, they had a certain way they treated black people."[42] That reality came home to the Southern University engineers one night just west of Huntsville in an Alabama town called Athens.

Even though "everything was more or less segregated" both inside Huntsville and out, there was a frighteningly noticeable difference even just a few miles away from Rocket City. As Watson put it, "We all said, 'Once you step out of Huntsville, you are in Alabama.'"[43] Athens was outside of Huntsville. What happened there went beyond the usual affronts and day-to-day discrimination they would face if they traveled south of Holmes Ave. That kind of stuff, Bourda said, the guys handled pretty well with humor. There was the time they went to a restaurant and "they said, 'We don't serve niggers,' and Morgan said, 'that's alright, we don't eat 'em.'" Another time, in a Huntsville bowling alley, George recalled, "You could eat, but you couldn't bowl. We went to bowl and the lady said, 'I'm sorry, you can't bowl.' Tommy said, 'Can I eat?' 'Yeah.' 'OK, then I'm gonna eat!'"

This time was different. As the four young well-dressed black men walked in the door, it was quiet, with only a waitress and a couple of customers at the counter. The NASA co-ops walked first to the "colored" side, which was closed. Then they turned.

"We're hungry," one said. The waitress replied, "Colored side's closed." One of the men said, "We know that. We were just there. We're not stupid."

"Look," she said, "You can't loll around."

"C'mon, we want to get hamburgers."

"She said we were loitering," Bourda recalled, and told them to get out. That was not going to happen this night. As Watson remembered, "We said a few things." Everyone got into everyone's face. There was some shouting. That is when the waitress pulled out the white trump card. She said she was going to call the police.

Bourda knew what that meant, or what it could mean. Watson and the others knew it, too. To a white cop in Alabama, Bourda was no NASA engineer. Watson was no Water Walker. They were dirt. They were less than dirt. And what was the value of dirt—not anything that anyone would miss. In fact, at the same time this was happening, three young men were missing in Philadelphia, Mississippi. When their bodies turned up a couple of weeks later, the FBI said the people who tortured, shot, and buried them worked for the Neshoba County's sheriff office. "Once you step out of Huntsville, you are in Alabama." They were outside of Huntsville. They did not wait to for the police to show up. "We were in two separate cars," Bourda said. "We left."

The View of What Was to Come

"We knew that on the social scene everything would be segregated," Watson said. But "on the other hand, on the work scene, that was the view of what was to come." NASA had truly placed the Southern University co-ops at the cutting edge of rocket technology at Marshall. Unlike the work the mathematicians were doing, which the black press had to puff up to make sound important, Watson and the others were actually engaged in what the public generically referred to at the time as "rocket science." Frank Williams helped design the ground-support equipment for the Apollo program.[44] Morgan Watson started out in the Quality Assurance Laboratory. There he worked "testing daily the various components of the space capsule," looking at its temperature and pressure and simulating how they would react in space along with "making sure that all of the screws were tightened, no wires were corroded or anything of that kind."

While at MSFC, Watson took some advanced courses to "understand the aspects of the engines more so than just the regular test work, which was routine." As a result, he left the Quality Assurance Laboratory and moved to propulsion, where he worked on testing the Saturn 1-B, the forerunner of the Saturn V, "all the way from the paper designs . . . to the static firing," where the engines actually roared to life. "I felt more comfortable there than I did in Quality Assurance because that was mostly electrical and electronic. So I had a chance to work within my major and to see how the things I learned in school actually applied in the space program." While in propulsion, he worked "in the heat shield area," developing "ceramic materials and various honeycomb shaped materials to help withstand heat and prevent rockets from exploding." He also participated in live firings. "The live firings were done right on site—at the Marshall Space Flight Center—way out in the back. It was a sight to see. The rocket is tied down and the engines are firing and it's quite an experience," he said. "A lot of noise too."[45]

Watson and Bourda said that NASA's work was not just technologically groundbreaking. The space program "was the high-tech industry of the '60s," Watson said. "Only the best and brightest could ever aspire to go there." It was, Bourda echoed, "the high-tech industry and only the chosen could go there."[46] Watson realized the effort NASA was making and he said, "We just praise NASA for its foresight to go and find African American engineers to train."

Regardless of how people felt about the SU-BR students being there, they caused a stir. "We were sources of curiosity by everybody," Watson said. African Americans at the facility "were proud of us because we were just blazing new territory for them to follow behind," while whites just "wanted to know who we were and where we came from." He remembered someone once coming up and asking if he and the others were visiting students from Africa. That is how far beyond the realm of possibility their experience seemed. A white Alabaman simply could not imagine that a group of young black men walking through a NASA facility in white shirts and ties would be Americans.

Watson and the other African American co-ops did not have the problems with their co-workers that Julius Montgomery had endured in Florida. "We had people from all backgrounds working in the space program," Watson said, including many people from northern universities "who had gone to school with black students before." He admitted that the aura of the space program scared him a little bit, and at first he doubted himself. "The students that they had from the University of Alabama and other universities in the co-op program—we always had in the back of our minds going to a black school: maybe they were getting a better grade of education than we were."[47] It turns out, Bourda said, the white students thought the same thing. "I'm light skinned," he pointed out, and so "I look white." As a result, "they thought I was one of the guys. Every now and then they'd slip up and say something and [only then would] they realize I was there." When this happened, it laid bare the prejudice embedded in an organization of exceptionalists who had invited in outsiders whose merit and character they questioned. Bourda said it was common to hear the white co-ops say resentful things about him and the other black students when they thought they were speaking in confidence. "We got a lot of 'em trying to come in," they would say. And more to the point, "I don't think they can cut the mustard."[48] These same problems would crop up again fourteen years later, when NASA first allowed women into the astronaut corps. At the time of their arrival, the agency did not have separate showers, dressing rooms, or toilet facilities for women at the astronaut training facility in Houston; women, like blacks fourteen years earlier, simply did not belong. NASA was accustomed to always doing things the same way. Change did not come easily.[49]

Whatever doubts NASA or the other co-op students had, Watson said his doubts evaporated rather quickly once he was able to talk with the white students about the quality of their education. He learned that at SU-BR, "We had much smaller classes; we were not taught by graduate students; we were taught by the professors themselves. We convinced ourselves that we had a great education and we were not arrogant by any means, but we were very, very proud of ourselves and we liked the opportunity to be able to show that we could perform." In fact, Watson said he had the opportunity to teach some of NASA's older hands some new tricks. "I was in Southern University's first computer programming class back in 1963 and when I got to NASA I was one of the few—including NASA employees—who was computer literate." Computer technology was in its infancy then and "as a result, they were sort of suspicious of the guys who could use a computer. They would still check everything by hand. After you ran the program you had to check it to make sure the computer was right." In this regard, he said, "we were pioneers in more ways than one because we brought skills that many of the old-timers at NASA didn't have."[50] In the end, it all worked out. Of the other co-ops, Bourda said, "They realized we're human just like they were."[51] Watson said, "We all got very good evaluations by our supervisors—even though they may have been apprehensive—and [they] felt that the experiment worked."[52]

Participation in Events Which Are Not Attended by Negroes

At the time Watson and Bourda were at NASA, the agency took the opportunity to attempt several other experiments, too, as it worked to push back against the backward racial policies of its home communities in the South. People in the Kennedy administration had always seen the space agency's administrator, James Webb, as one of Lyndon Johnson's guys. That made Webb's situation tenuous during the days when the White House staff schemed behind the vice president's back and called him names like "Rufus Cornpone" behind closed doors.[53] After Johnson ascended to the White House, agency heads that had backed him during the hard times felt much more secure. In 1964, Webb, for his part, was able to demonstrate his fealty to Johnson and to Johnson's policies on integration and civil rights. That position helped Webb, no doubt after an incident that year that had the president furious over NASA's actions in one of its host communities.

Because NASA was spread out all over the country, James Webb often had to take on a different persona and—sometimes—a different set of priorities, depending on which center he was standing up to highlight or protect. That helps to explain his behavior in the episode that annoyed President Johnson. A key element of NASA's rendezvous with the culture of the South involved the symbolic question of whether or not NASA staff people could attend meetings or speak to groups that discriminated. The agency wrote a policy against speaking at segregated meetings in 1963, and it had a form letter that allowed it to turn down invitations. Those worked as far as they went, but, when the person making the invitation was a VIP, circumstances forced agency people to vacillate between responsibility and reality.

John Stennis of Mississippi was one of the most powerful men in the United States Senate and such an important backer of the space program that NASA eventually named its Mississippi Test Facility after him. But in 1964, Stennis was furious over NASA's policy on public speaking and public appearances. Webb had scheduled a speech to the Jackson, Mississippi, Chamber of Commerce. But two African Americans—members of the Student Nonviolent Coordinating Committee—heard about his talk and tried to get tickets. The chamber, which discriminated against blacks, turned them down. NASA policy said that meant Webb would have to cancel. Of course, if he did, Senator Stennis would be furious, and Webb would have to worry that Stennis would hold it against him and NASA the next time the budget came around.

The president first learned about this situation when he got a call from aide Walter Jenkins, who had just gotten off the phone with Webb's executive assistant, Colonel Lawrence W. Vogel. Jenkins briefed the president about the situation in Mississippi and told him that Webb intended to go ahead and make the speech "unless we told him he couldn't—which I thought was a heck of a way to put it." Jenkins told the president that he had reminded Vogel about the policy on speaking at segregated meetings and that Vogel had pushed back, telling him, "Sen. Stennis is one of our main supporters and we talked to him and he says, 'Oh that's just a little group nobody pays attention to.'" Jenkins said he was bothering the president only because the "Student Non-violating Committee," as he called them, was planning to stand outside the venue and "have demonstrations and placards and so forth. I would hate to bother you with this but I want to make sure the line I'm taking is the right line," he said.

The president was livid. "Webb's bound to have more sense than that," Johnson said, and ordered Jenkins tell him so. Call Vogel back, the president said, and "just tell him, 'The President said Webb's bound to have more judgment than to go to a segregated meeting when they'll being parading.'" The space program is "too important to have it segregated," he said.[54]

Webb apparently got the message because later in the day, an infuriated Stennis called the White House to lodge his own complaint with the commander-in-chief. As with the July 18, 1963, meeting between Johnson and Robert Kennedy, reading this exchange is not as effective as hearing it. The recording is a classic example of what came to be known as "the Johnson Treatment"—a verbal assault on those threatening *lèse-majesté* that *New York Times* columnist Tom Wicker said usually left the offender "intimidated, unnerved, [and] reduced to a sort of nothingness" in the great man's presence.[55] On the tape, the listener can hear Stennis's expectations and his ego deflate simultaneously as the call goes on. Early on, he is strutting. By the end, he is groveling, as he seems to back out of the room, begging pardon. "John, don't get me in on Chamber of Commerce meetings," the president said, chuckling derisively. "I got a bunch of folks here." In other words, how dare you bother the president of the United States with this kind of petty nonsense?

> STENNIS: I know that
> JOHNSON: And I just can't do that, my friend. That's something that
> STENNIS: Well—
> JOHNSON: I don't want to even know about if I can avoid it because it will have a lot of complications.

STENNIS: Alright. I didn't know I—

JOHNSON: You just forget you talked to me. They got a rule here that nobody in authority speaks at any segregated meetin'.

STENNIS: Well I would—I didn't want to involve you. But I'm not gonna say I called you.[56]

The fullness of the agency's dilemma is visible in this one episode. The nation has a law. NASA has a policy that complies with the law. An agency patron lives where people do not want to comply with the law and asks the administrator to break the policy and break the law. NASA calls the highest authority and asks for permission to break the law. The president says no. NASA tells the patron no, and the patron gets upset. The *Informer*, a black newspaper in Houston, printed an article about black groups asking Webb to cancel his speech in Jackson. A few months later, the Johnson administration issued formal guidelines for federal officials regarding speaking engagements and conferences.[57]

The Stennis episode was not the only time NASA ran into a problem with its policy on segregated meetings. Houston's Petroleum Club was located on the top floor of Humble Oil's forty-four-story headquarters. The Manned Spacecraft Center (MSC) sponsored an educators' conference and seminar in the fall, and part of a NASA group invited to the club for after-dinner drinks included two African American educators. The club refused the two men admission, and a NASA public affairs officer appealed to the club manager to no avail. Though members of the conference board at the club met and apologized to their guests, Dr. Robert Gilruth, MSC director, wrote the Petroleum Club and explained government policy on segregated functions and venues. He informed the club of NASA's embarrassment and said that the agency would no longer use the club in any official capacity. When Time, Inc., leased the Petroleum Club for a cocktail party in March of 1965 to introduce a new reporter, Ben Cate, to Houston, NASA officials received their invitations through the MSC, making the event an official function. When Gilruth learned that the Petroleum Club was the site, he informed the NASA guest list that they were not to attend; thus thirty NASA employees, including seventeen astronauts, were absent from the party.[58]

Once again, the constant rendezvous with the cultural strictures of the South forced NASA to decide whether to comply with the law or bend to local custom. Webb was clearly trying to go around the rules in Mississippi; hence Johnson's admonition that he "have more judgment." Gilruth rightly chided the Petroleum Club, so one might say NASA was learning—but then there is the agency's behavior in Alabama at this time.

On March 4, 1964, Wernher von Braun's top deputy at the MSFC, Harry Gorman, wrote to Al Hodgson, the agency's racial understanding czar. If Gilruth was having trouble enforcing new civil rights laws in Texas, imagine how difficult it was for von Braun in Alabama. That is why Gorman was writing. The letter asked Hodgson for a blanket amnesty from the speaking ban in Alabama. "There has been an increasing interest on the part of various groups, including members of NASA headquarters staff," the letter said, "to implement a policy which would require that employees of NASA avoid speaking at assembled meetings, symposia, and the like where, as a matter of custom, minority groups would not be accepted in the audience."

The letter went on to enumerate all of the steps that the center had taken up to that point to try to increase the number of African Americans working there. It then said that although "admittedly, we have not been as successful as both you and we would like in employing qualified Negroes," people at the center felt that they had created a climate of racial moderation and understanding in Huntsville and promised that "this affirmative policy will continue in the future." Then, abruptly, the letter turned on a dime. "The conduct of NASA official business in the states of Alabama, Mississippi and Louisiana," over which the Marshall Space Flight Center had jurisdiction, "frequently requires the participation of our people in events which are not attended by Negroes. Such events include meetings and conferences with business and state officials and seminars of the scientific and technical nature." These meetings are vitally important, the letter suggests, and then asks that "if a published policy is considered necessary, I think we should be most careful that our ability to handle normal business is not hampered or rendered ineffective."[59]

Later in the year, von Braun told U.S. senator John Sparkman of Alabama that he would only be able to speak at the annual convention of the Alabama Homebuilders "provided that it is a non-segregated group," so it is clear that Hodgson or someone at headquarters had denied their request.[60] But it is difficult to tell overall whether NASA's stand on this issue was closer to Webb's, Gorman's, or Gilruth's.

As in the Stennis phone call, Lyndon Johnson demonstrated, in large ways and small, a heartfelt commitment to creating equality among the races. A close reading of PCEEO transcripts does not show Johnson to be a venal politician engaged in a manipulation of numbers designed to win votes. Instead, he comes across as a man who knew poverty and racism first-hand; who felt that he understood their interconnectedness and their corrosive effects; and who truly believed America would be a better place if he could create and implement the means to end them. Following the riots in Birmingham and Selma, and George Wallace standing in the schoolhouse door, "the emotional and political conclusions were unambiguous: the civil rights problem was exploding intolerably as a Southern problem."[61] Discrimination, racial separation, and the forceful efforts to keep them in place were a concern throughout the region. With perspective and in that context, a particularly significant action that Webb took on Johnson's behalf seems not well thought out and somewhat difficult to comprehend. Nevertheless, it applied a dual set of pressures on Alabama that the president no doubt noticed and appreciated.

A "Hate [the] Federal Government Doctrine"

The action Webb took, shortly before the 1964 presidential election, was either the firmest, most principled stand taken by NASA leadership against George Wallace, or it was Webb going out of his way to use federal power to sway the election to his patron. His real purpose is difficult to reconstruct so many years later. But the crisis was legitimate and it caused considerable panic in Huntsville for some time.

The incident began with a misunderstanding and snowballed from there. Webb spoke in Montgomery to the Alabama Chamber of Commerce in October. During the question-and-answer session after the speech, he "stated frankly that we were having trouble recruiting scientists, engineers and managers and that this was a matter which I believed to be of importance to the businessmen as well as to the management of NASA."[62] News coverage of the speech gave Louisiana congressman Hale Boggs an idea. He called Webb and told him that if the agency was having trouble with the recruiting of people to Alabama, he should move the MSFC. Not surprisingly, his solution was to move it to Louisiana. Congressman Boggs liked his idea so much that after he finished talking with Webb, he appears to have hung up and called the *New York Times* to tell their NASA correspondent John Finney that "hundreds" of top MSFC officials would be moving to NASA's Michoud Facility in New Orleans. Finney wrote an article that further muddied the waters, saying that "the National Aeronautics and Space Administration announced today it was considering the transfer of some of its top scientists and executives" out of Alabama, even though NASA had said no such thing, and his article contained no such announcement. In checking out Boggs' claim, Finney appears to have contacted Webb and, after speaking with him, realized that Boggs had gotten it wrong. Though he stated that fact in the middle of his article, he did not alter the beginning or the end of the story. Thus, he ended up leaving the impression that NASA was, was not, and was planning to move its people out of Alabama.[63]

The ramifications for Huntsville would have been devastating. If NASA moved from Alabama to Louisiana, it would mean thousands of jobs lost—NASA jobs, of course, and those of Alabama-based NASA contractors. But it would also hurt teachers, home and road builders, store owners, and, as von Braun liked to say in speeches, "the Laundromat operator, the waitress in the motel restaurant, the entrepreneur at the parkway hamburger stand"—those who constituted Huntsville's economic ecosystem.[64] The rumor of a reduced NASA presence panicked the business community. Jim Dunn, president of the Huntsville Home Builders Association, telegraphed the White House to report that financial institutions had stopped construction loans. He predicted labor and personnel problems as a result.[65]

The confusion over what Webb said or did not say shadowed him when he spoke to a different group of businessmen in Huntsville several days later. In a meeting with concerned members of the Huntsville Industrial Expansion Committee on October 29, Webb laid out the reason why NASA used Huntsville for the job of "managing the contractors who assemble, test and launch giant boosters," adding that "we want to keep this arrangement." However, he also told them that "it would be good business" to move "part" of this work to Louisiana, or possibly California. He suggested that "attracting senior executives from industry to the Huntsville installation" was posing a problem for

NASA, saying that thirty people had turned down the opportunity to move there, though he did not say why. He also told the executives that the amount of work that remained in Alabama "will directly depend on our ability to improve" recruitment. "To attract the best, we must offer them an environment in which they will want to live and work," Webb said. This was an argument he and von Braun had used many times in the past in speeches promoting more spending on schools, libraries, roads, and universities in Alabama.[66]

When Webb met with the press after the meeting, the reporters downplayed, skipped, or ignored these nuances entirely. The United Press International moved a story that was carried in many black newspapers quoting Webb as saying Alabama needed to improve its "image," and adding that "he attributed this 'image' largely to racial segregation." The article said Webb told reporters that "the question of the race issue is raised in recruiting conferences." A column in the *Washington Star* reported that Webb told reporter Robinson J. Willard of the *Huntsville Times* that there was a "hate [the] federal government doctrine" in Alabama. The paper said he blamed that doctrine for the agency's "very real difficulty" with recruiting.[67] Webb also called Alabama "the state that is the least appreciative for all the federal space dollars being poured into the Southeast."[68]

The charge that thirty executives had rejected jobs in Alabama because of the state image on race relations prompted an editorial in the *Huntsville News*. The paper challenged Webb to supply the names of the thirty executives. Since Webb "continues to talk in vague generalities about our bad image," the *News* editorial said, "perhaps he would do his own NASA and the state a service if he would give to the Huntsville Industrial Expansion Committee the names of the 30 executives who 'would not move to this state to take positions.'" A NASA spokesman, when asked, would not reveal the names.[69]

Robert B. Young fit the storyline. Young moved to Huntsville from Aerojet Corporation in 1963 to take over the agency's new industrial operations office, which dealt with contractors. He left NASA after a year, "mostly because his family refused to move in the face of mediocre schools and the state's reputation for bigotry and violence."[70] The Marshall Center periodically used the state's racial situation as an excuse for not being able to find more African American workers. Similarly, the manager of the Huntsville division of Northrop Space Laboratories wrote a letter to Governor Wallace right before the schoolhouse door incident where he said recruits feared a "racial mess" in Alabama.[71] In light of all that, Webb may have been taking a principled stand against racism by talking about his recruiting problems in Alabama.

However, his invocation of a "hate [the] federal government doctrine" might have had another motive. A presidential election loomed while all of this was going on. There is some suggestion that NASA's presence in the South blunted local displeasure about the Civil Rights Act and the president who signed it into law, at least in some places. As the 1964 presidential election approached, William Ellis of the *Nation* wrote of a Florida businessman who told him that when it came to the civil rights bill, "I don't like it, and I guess that I'm lined up with [Republican Arizona senator Barry] Goldwater on that point." But the man said that if he had a choice of "letting colored people sit next to me in a restaurant or having Cape Kennedy closed up, well, brother, you can bet I'll let them sit. Not only that—I'll pass them the sugar."[72] Frankly, though, there were not enough people of that sentiment to forestall Johnson's prediction that, with the passage of the Civil Rights Act, the Democrats would lose the South for a generation.[73]

What if the election was close, many wondered; would the loss of Alabama play a role in Johnson's defeat? This threat to pull thousands of federal jobs out of the state—was this payback for the state leaving Johnson off the ballot? Was it a threat that he would lift days before the election so thankful northern Alabamians might vote Johnson/ Humphrey? That kind of speculation filled the press at the time.

"Alabama, the only state in the union without President Johnson's name on the ballot, may win electoral votes for Goldwater but lose space prestige and payroll to Louisiana," *Aviation Week* said.[74] "Democratic electors are not on the ballot in Alabama, and the state therefore is generally conceded to Senator Barry Goldwater," the *New York Times* said in its article about NASA's "announcement," which it said "could help sway voters over into the Democratic column."[75] Goldwater's people wanted to make sure the press placed Webb's announcement only in the most negative and partisan light. Republican National Committee chairman Dean Burch challenged President Johnson to "deny if he can" that Webb was "used as a political messenger to attempt to browbeat the people of Alabama into supporting the Johnson-Humphrey ticket."[76] To bolster the charge that this was all just a partisan exercise, *Space Business Daily* pointed out the self-evident hypocrisy of those trying to frame Webb's decision as having anything to do with race relations. In addition to mentioning a possible move to Louisiana or California, right after he left Huntsville, Webb told a business group in Little Rock, Arkansas, that "the possibility [exists] that many varied research, development and production activities will gravitate in this direction." The paper declared, "The whole nasty mess" pitted race relations in Alabama "against those of Louisiana, or Texas, or Mississippi, or Florida." Either Webb was comparing the race relations between states or he was saying that NASA was going to have to move all of its facilities out of the South. As the magazine pointed out, "It seems only a very short time ago that Arkansas and Little Rock were having their share of embarrassing race relations." The paper suggested that what Webb was saying about Alabama was not about Alabama at all, but was instead about George Wallace, who had worked overtly, and some could say aggressively, for Lyndon Johnson's defeat. The article strongly suggested that it was just plain silly to suggest that race relations are any better in Arkansas, Florida, or any of these other places than they were in Alabama, thereby leaving only one conclusion: that Webb was doing this for purely political reasons.[77]

Johnson would later defeat Goldwater in a landslide. Of the southern states NASA called home, Johnson carried Florida and Texas, but he lost Louisiana and Goldwater crushed him in Mississippi. Goldwater also won Alabama; but of course no one knew that in October 1964.[78]

A Man Whose Allegiance Is Ruled by Expedience

After the election was over, Webb left it up to Wernher von Braun, of all people, to suggest otherwise. By the mid-1960s, news about von Braun's SS membership had begun "to seep into the Western media," as had news that concentration camp prisoners working at Peenemünde had been beaten, starved, and tortured.[79] Nonetheless, there was no one who had more influence in Huntsville. Despite his flaws as a messenger, von Braun was sent out to make the case that Webb's threat to move NASA *was* about race relations *and* about the community's reputation, and that if people wanted to maintain the professional status quo, then the racial status quo was going to have to change.

When the agency announced the transfer of a hundred-member space launching technical team from MSFC to Cape Kennedy in Florida, people in Huntsville saw it as a harbinger of future moves out of Alabama.[80] With the community on edge, von Braun prepared a speech that he knew needed to "walk a tightrope between helping the Huntsville credit market and not crossing up Mr. Webb."[81]

In that speech, to the Huntsville Chamber of Commerce, von Braun requested that everyone in the room ask him- or herself, "Are you doing everything in your power to strive for fair employment and improvement of racial relations in our city?"[82] Webb, he said, "was right, you know. It has been difficult to get top-level people to come to Alabama and to Huntsville," though he suggested the area's reputation was an unfair one. People did not fear moving to "the Huntsville we know— but to a place whose image has suffered by what has been said about it and what has been written about it nationally." Nevertheless, he acknowledged that "we should all admit this fact: Alabama's image is marred by civil rights incidents and statements." He went on to talk about the Huntsville contractors group, saying that "I know that it took more than a little courage for this association to undertake such leadership," and he urged everyone to become familiar with the equal employment opportunity section of the Civil Rights Act of 1964 and "the rights afforded and the obligations imposed by its provisions."[83]

This speech came at a time when the German rocketeer was acting as "a cautious but important voice for integration and racial moderation in Alabama."[84] Where only two months earlier he could address the Alabama State Homebuilders Association and never mention civil rights, now things had changed.[85] In December, von Braun headed to Fairfield, Alabama, to speak at the groundbreaking for the science building at Miles College, the campus that had served as the intellectual nerve center for the black community during the Birmingham unrest. Miles College led boycotts of Birmingham businesses and organized important voter registration drives, so when von Braun stepped on stage with the school's administrators, it made national news in the black press.[86] "The nation's foremost space scientist, Dr. Wernher von Braun, broke ground at Miles College in Birmingham for a $450,000 physics building," *Jet* magazine said, quoting von Braun as saying that "Miles College is serving notice that it plans to offer the community an opportunity to keep pace in this new technological world of ours."

The *Chicago Defender* gushed about the "eminent space scientist," who it said was known as "Mr. 'Space,'" going down to Miles. The paper showed him speaking to "an audience of 1,000 people" from a podium emblazoned with the college's name.[87] L. C. McMillan, the executive director of AHAC, the Huntsville contractors group, hailed the Miles College speech as being "most encouraging to those of us concerned with the problems of motivation, education and equal opportunities of minorities." Von Braun's message "regarding equal opportunities and the importance of education hit its mark or at least, it is well on course!" In addition, McMillan said that local "minority leaders" had come to him several times in the past few days wanting "to discuss your speech and its implications."[88]

Later, von Braun spoke to the Huntsville Rotary about the importance of voting rights for all Americans. Brown Engineering's Milton Cummings, AHAC's original leader, praised the speech. In contrast, Linton Crook, a retired U.S. diplomat living in Alabama, told von Braun that "if giving the right to vote to every man, no matter how simple or uninstructed or limited by nature, were the answer to the problem, we should, indeed, have very little to worry about."[89]

Wallace and Huntsville, Part 2

As 1964 turned to 1965, NASA leaders believed they had the moral authority to push Wallace even harder. In March, Alabama state troopers brutally beat a group of African Americans protesting voting restrictions while they knelt and prayed on the Edmund Pettis Bridge in Selma. President Johnson met with Wallace at the White House on March 13, and the president made clear his revulsion at the violence and his determination to prevent its reoccurrence. Two days after the Wallace meeting, Johnson proposed what became the Voting Rights Act of 1965 to a joint session of Congress.[90]

Two months after that, Wallace organized a road trip for small-town newspaper editors. The governor's idea was to show them "the real Alabama" rather than the "distorted" view they got from the mainstream media. While the turnout was small—twenty-seven editors came from tiny communities in sixteen states—twenty-one reporters also attended the event from major publications like *Time* magazine and the *New York Times*. They tagged along to write about just which sights Wallace chose to show the small-town editors while they were in his state.[91]

The trip came during yet another fraught time in Alabama. In addition to the continuing fallout from the Selma protests, Wallace was once again in a battle with the University of Alabama and the federal government over integration, this time involving the university hospital. As of January, the Civil Rights Act prohibited discrimination in hospitals receiving federal funds, and at the time of Wallace's road trip, federal inspectors were in Birmingham trying to decide whether University Hospital was in compliance.[92] Vivian Malone, one of the students whose enrollment Wallace tried to block, was set to graduate in May, bringing the stand in the schoolhouse door back to national attention.

Webb had recently testified before Congress, along with University of Alabama president Frank Rose, and had hit Wallace "pretty hard" there, including the issue of discrimination at the hospital.[93] At the hearing, Webb said he was "anxious to have the standards raised by somebody who intends to go forward instead of following Wallace backwards."[94] Webb's high-profile position got Wallace's attention. "Governor Wallace has apparently tried to make a big issue out of the thing to let it be blasted to the world," von Braun told the head of the Army Ballistic Missile Agency, General John G. Zierdt.[95] As part of his plan, Wallace decided to add a new stop to his "real Alabama" road show; a June 8 visit to the Marshall Space Flight Center. Wallace loved to attach himself to NASA's success. In fact, at that very moment, the legislature was contemplating funding a new facility at a decommissioned airport that von Braun learned would be called the George Wallace Space Museum. When Webb learned Wallace planned to visit Marshall, he decided to go to Huntsville himself. "I'll not allow for Wallace to come in and take credit for what has been done," Webb said. According to Webb, Wallace was "a dangerous man" who would "stop at nothing."[96]

Planning together, Webb and von Braun decided it would be important to show that the army was on board in their effort to stop Wallace. "I think it would be of utmost importance that the Army plays exactly like NASA," von Braun said. "It should be perfectly plain that this is not just NASA, but also the government." Webb suggested that von Braun call General Zierdt to warn him about the visit and to tell him "that Secretary of the Army [Stephen] Ayles [*sic*] is in on the act." Von Braun tried to one up Webb on that one by asking, "Do you think the Secretary of the Army would come personally?" Webb said he did not think so, "because he has to fight a war in Vietnam." Webb said that he had spoken with the secretary of the army and he agreed that Webb should make sure Wallace could not "make a single move without meeting me head-on." To that, von Braun joked, "He may cancel his trip if he knows that you'll meet him in the schoolhouse door."[97]

Wallace was doing some strategizing on his own. Anticipating that Webb and von Braun might give him a problem if he arrived alone with just the reporters, Wallace upped the ante. By the time von Braun called NASA executive officer Lawrence Vogel later in the day, the delegation coming with Wallace had swelled. In addition to the reporters, 120 members of the Alabama legislature would be showing up on the morning of June 8.

This was no longer some road show for small-town newspaper editors. It was a major national event. George Wallace would be there along with most of the Alabama state legislature. The head of NASA would be there, as well as Wernher von Braun and twenty-one members of the national press. This story would be front-page national news the next day. If Webb and von Braun stuck to their original plan—a drop-by and a walk-through—Wallace would be the center of attention. He would be the one framing the event. For NASA to change the narrative, something was needed to turn the spotlight back on Webb. There were few things in 1965 bigger than George Wallace's mouth and ego, but fortunately, NASA had one—the Saturn V rocket. Von Braun made a decision: after staff had briefed the legislators and given them a tour of MSFC, Governor Wallace would show up with the reporters, and NASA would put on an explosive show. The assembled guests would receive the rare treat of witnessing a static test firing of the Saturn V's first stage. This was the equivalent of a Moon rocket launch with all the suspense, all the pomp, the countdown, and all the noise, just without the launch.

This was now a NASA show. Wallace had called a press conference for 9:30 that evening, and, as far as von Braun cared, he could have it because it would not be the next day's story. NASA would corral Wallace, the legislators, and the reporters into the visitors' area. There would be the countdown. Then there would be the explosion. When it was over, "we'll have them in the [visitors'] area for safety reasons," von Braun said. And that was when he and Webb would speak. The news took Colonel Vogel aback. "You'll have the full firing of the five engines?" he asked. Yes, von Braun told him, "we had three such tests previously but this is the longest one we ever had."[98]

On June 8 in Huntsville, as the delegations arrived, they were met by picketers carrying placards that said things like "look under the rug."[99] As was often the case with a NASA press event, things went off exactly as the agency wanted them. The *New York Times* reported that the delegation received "a lecture." The paper said that "both James E. Webb, the NASA administrator, and Dr. Wernher von Braun, director of the Rocket Development Center, obliquely but clearly warned Mr. Wallace and his segregationist followers in the Alabama legislature that the state might not achieve its promise of industrial expansion under the governor's policy of 'segregation forever.'"

Neither Webb nor von Braun referred to race in any way. Both largely mentioned the same types of things that they had been telling the legislature and chambers of commerce for the past ten years. Von Braun said Alabama needed to improve its education and other types of amenities that would "keep our best people from leaving the state" because it was unable "to offer them the same opportunities here that they can find in any other state."[100] Webb said, "The size and importance of [NASA] operations in Alabama require us to add our support to the efforts of forward-looking and fair-minded leaders of the state in approaching the problems we all face." The phrase that the reporters said referred to race came when von Braun said, "The era belongs to those who can shed the shackles of the past." The *New York Times* reporter took the word "shackles" to be a reference to slavery. Although the reporters had to read between the lines, and possibly indulge in their own wishful thinking to find opposition to Wallace's policies from von Braun, another official host on the tour was much more forthcoming. The *Times* article said that A. L. Johnson, president of the First National Bank of Decatur, rode on the bus with the reporters and editors and criticized Governor Wallace during the ride between the town of Coleman and his hometown. Johnson called Wallace "a fine Christian gentleman," adding, "I don't think there's anything wrong with his administration except his racial policies." Johnson did not make these remarks in a private conversation with one reporter in the quiet of the bus ride. He spoke over the bus public address system, standing next to Governor Wallace's press secretary, Bill Jones.[101]

Wallace held his press conference that night. The *Washington Post* said he talked at length about a "Communist conspiracy" at the root of the civil rights movement. Wallace scheduled four press conferences during the tour. At the press conference in Huntsville, which was the third, the *Post* said that the editors refused to participate. Wallace had seen to it that everyone who could stick around after the rocket test stuck around. As a result, there were two hundred or so state legislators and local leaders present, with the small-town editors sprinkled in among them. In the *Post* account, one of the editors, a man from Palo Alto, California, announced "that he did not intend to participate in a press conference with so many non-press people present." The editor from the newspaper in Beaver Falls, Pennsylvania, stood up to say he agreed, though he did use the opportunity to take issue with the whole premise of the road trip in the first place. "The thing that galls me," he said, "is that we are all branded as distorters. Myself, I resent it. I have pride in my profession." He added, "There are more politicians in jail than newspaper men," which the *Post* said got a laugh. With that, the reporter said Wallace shrugged his shoulders, gave "a chipper smile," and ended the press conference. Wallace had scheduled a fourth press conference for later in Mobile. He canceled that one outright.

George Wallace did not return to Huntsville, at least not for another public event during the remainder of NASA's mission to the Moon. Apparently he got the message Webb and von Braun intended to send. Any political capital he gained by standing next to NASA was going to come with a heaping portion of disdain. The calculus was clear, so Wallace kept his distance. NASA's rebuke of Wallace was an important step in the cause of race relations, but it was one the agency would get no credit for in the African American community.

"We Wanted Something Different"

As the 1960s pressed on, many African Americans realized their marches and protests for equal opportunity and access had fallen short of their expectations. While they may have destroyed segregation, racism and poverty were still firmly in place.[102] With that, the attitude toward the space program in the black press changed, too. No longer were articles about gaining access to NASA jobs; instead, people expressed opinions like those of Whitney Young, who had once seen Ed Dwight as the engine that would drive black student achievement. "For the poor," Young wrote in 1969, "the moon shot seems just another stunt." It was "a circus act. A marvelous trick that leaves their poverty untouched. It will cost thirty-five billion dollars to put two men on the moon," Young pointed out. "It would take ten billion dollars to lift every poor person in this country above the official poverty standard this year. Something is wrong somewhere."[103] A survey of the black press at the time of the Apollo 11 launch suggested that those thoughts represented African American opinion at large.[104] If there was ever a poll conducted that confirms or denies this, it is difficult to find.

We know for sure that among those African Americans who worked in the space program, the opinions skew completely in the other direction, just as you might imagine they would. They certainly doubted whether the space program would achieve any of the things Kennedy and Johnson hoped for, like addressing racism, alleviating poverty, or changing the face of the nation's technology workforce. But these people had spent years going to work every day, working hard and drawing salaries from a mission to put the first human being—an American—on the Moon. The pinnacle of American technological victory in the Cold War with the Soviet Union happened because of them, too; there is a united understanding among the space program's African American pioneers about this fact. Some of pioneers said it was tough to grasp the magnitude of what was going on in the moment, but that ended once the rockets lifted off. As James Jennings said, "At that time, you were just trying to get your little piece right so I didn't realize what it was. But I think, when I realize the enormity of it was when we saw that big Saturn V lift off heading for the Moon and then you just were overwhelmed with the little part that you played in making this enormous thing happen."[105]

The pioneers also understood their own significance—what their presence within the space program meant to race relations in America. Take Morgan Watson, for example, who worked on the launchpad for the Saturn V rocket that took human beings to the Moon. His and George Bourda's participation was beyond symbolic. Their participation was a tangible demonstration of NASA's efforts to effect change in the South by bringing its host communities in line with federal civil rights goals. It is also an example of Lyndon Johnson's experiment coming to fruition—creating a new class of professions for African Americans to aspire to and thrive in. When they were young, Bourda and Watson knew the odds were against them, but they also saw that times were changing and understood that this time, when "we wanted something different," as Bourda put it, the chance to grab it was actually in reach. These men had no interest in being teachers, like generations of other educated blacks. "The teacher salaries were lower. I didn't see how I could deal with teaching," Bourda said. He was good at science and math and, he said, "we had advisors. One said, 'Why don't you try engineering?'"

As Johnson hoped it would, the space program "opened the door" for Bourda and the others. "It helped us a lot! We were determined and we had the ability to strive," Bourda said. Strive they did; and achieve too, and not just at NASA, Watson said. "Also NASA contractors, because Boeing and Chrysler and all of those were very, very anxious to hire black engineers and other professionals at that time." Watson said, "When I graduated, I didn't work for NASA directly. I worked for Chrysler, which was one of the NASA contractors. They made the first stage of the moon rocket—the booster rocket that took men to the moon." Later on, he said, "I had the opportunity to work at McDonnell Douglas and they made some of the components for the space program as well." This was important, he said, not only for himself but also for America. "By showing that there were black professionals, black engineers and others—computer scientists and others—that could do that, you proved the fact that people were available that could do it." Today, he said, "kids know, the sky is the limit." As George Bourda said, "Since I went into engineering—in my family we have eight engineers in the family. I'm proud to be the first one. The kids—they saw what I did and they copied."[106]

It is worth remembering that it was in the middle of a cotton field that Morgan Watson first began to think about becoming an engineer. He was a young man, stooping in the hot sun, ignoring the taunts of the white kids passing by, putting it all aside and vowing that he too would achieve the American Dream. At the time, Watson was just living his everyday life. He had no knowledge of the plans and projects that people in the halls of power in Washington, DC, were conceiving, and he certainly never imagined that they would ever touch him. "It was very difficult for me as a boy picking cotton in ninth grade/10th grade/11th grade to even visualize going to NASA," Watson said. "I knew what NASA was and I knew about President Kennedy's idea of putting a man on the Moon. But to think that I could work there!?"[107] Unimaginable then, but of course it did happen. For Morgan Watson at least, the wildest dreams of those who had a dream came true. And they did so because of the space program, which literally took him from the cotton fields to the launchpad.

Conclusion

James Jennings, Clyde Foster's young mentee from Alabama A&M College's first computer class, ended up spending his entire career at NASA. Over the course of his thirty-five years there, Jennings had the opportunity to be part of the agency's Council on Equal Employment Opportunity. NASA populated the council with deputy center directors from across the agency, the equal opportunity officers from all of those centers, and four selected employees from across the agency. Jennings started out as one of the four employees and by the early 2000s was council chair. This experience gave him a remarkable perspective on the EEO issue at NASA, enabling him to see how it has changed and how it has not. Considering Jennings' background, the length of his tenure at the agency, and his continued interest in the hiring and promotion of African Americans, his view is unique and valuable. His conclusion, therefore, is somewhat depressing. Overall, he said, "we haven't made much progress." In fact, he said, "the same things that we were talking about in the early '70s about getting blacks involved in the program or hiring blacks—they were the same problems that we have today."[1]

That assessment contrasts with one made by someone else who had a firm grasp on the issue. After President Kennedy's death, the John F. Kennedy Library Project began an oral history program designed to "recapture the experiences and impressions of those who served during the eventful one thousand days of the Kennedy presidency."[2] John F. Stewart was the project's chief. In 1967, he sat down with Hobart Taylor Jr., who was Lyndon Johnson's principal ally on the President's Commission on Equal Employment Opportunity (PCEEO), for a conversation about Kennedy's effort to open the federal workforce to

African Americans. Considering how little is written about the role of NASA in the PCEEO's project, their exchange is striking. Stewart began by asking if Taylor remembered, from his work with the PCEEO, which agencies "were outstandingly successful, and which were quite opposite?" The first agency Taylor mentioned was NASA. The significant exchange came next. NASA, he said,

> worked at it under very difficult circumstances because their facilities, and their—
> STEWART: Type of people—the occupations involved were quite different.
> TAYLOR: That's right. And also the location of all the NASA facilities are in the South . . . where they had tremendous problems of housing, and everything else of that kind, too, you see? But I think that the people of NASA were very intelligent.[3]

Two things are notable. First, Taylor pointed out the particular difficulties NASA had because it was in the South. As the man who helped institute Johnson's plan to change that region of the country, Taylor's understanding of NASA's struggles with the South's peculiar institutions was acute, and he understood, as Johnson did, why that presence—a federal presence and one of futuristic, technological excellence—mattered. If the South was to be changed, new jobs in new fields had to replace the archaic status quo. And if the South was to be changed, federal intervention was imperative.

The second notable thing was more subtle. Taylor shot down Stewart's suggestion that NASA jobs were too sophisticated for African Americans to hold. In a conversation about federal jobs, Stewart walked in assuming that at NASA, the "type of people—the occupations involved were quite different." At NASA, he suggested, the challenge was not filling the kinds of jobs they had at the post office or the Federal Highway Administration. At NASA, it must have been impossible to find black people for jobs that were "quite different." Taylor refuted that contention immediately, shooting down the idea that NASA jobs were too sophisticated.

As a member of the African American upper class, Taylor read the black press and saw dozens of stories about technicians, mathematicians, engineers, and pilots who went to work for NASA. No one can know whether he read stories about Frank Crossley inventing his new class of titanium alloys; about Morgan Watson and George Bourda going to Huntsville to work as part of the nation's technological elite; or about Clyde Foster's role in starting the Eastern Conference of Black Mayors. There is no doubt, however, that he came away from the black press reinforced with a feeling that allowed him to tell Stewart that "the people of NASA were very intelligent"—and he meant all the people of NASA: black and white.

That is a vitally important distinction in considering the role NASA played in the Kennedy administration's push for equality and the Johnson administration's drive for it. Equality is not just about access. For people to be equal, they need to perceive each other as equal. For many Americans, especially in the South, however, it might have been less difficult to accept a Chinese American or Mexican American co-worker as equal than to accept an African American. NASA played a role in changing that. As Morgan Watson said—and it bears repeating—NASA "helped to break the walls down. It helped change people's perception of black people in the South."[4]

It did this in both overt and subtle ways. The Marshall Space Flight Center in particular went out of its way to promote its African American employees by putting their photographs in the house newspaper, the *Marshall Star*. The center cited one African American worker for sustained superior performance in a film that it produced to promote an incentive awards program.[5] It is easy to see this as tokenism, but consider the culture in which they took this action. If, as Jack Spielman, the EEO hearing officer, found, "discrimination continue[d] to exist" at MSFC, then lauding an African American worker for his sustained superior performance was no small thing. In the same way, the dustup over employees attending segregated meetings and the restrictions on staying only in integrated hotels contributed to a sense that, at NASA centers, everyone was allowed to share in the sense of exceptionalism.

As Hobart Taylor Jr. said in the JFK library oral history interview, what was most important was the location of NASA's facilities and the inherent challenges that came along with that location. There were practical considerations beyond regional economics that brought NASA to the South and kept it there, but many in Washington, DC, knew or hoped they knew that having it there might change everything.

One reason why there has never been a detailed exploration of NASA's role in civil rights is that historians perceive the two issues as being completely separate. Quite the contrary; they were connected. Lyndon Johnson certainly saw them that way and, as chairman of the National Space Council, the PCEEO, and then as president of United States, he was in a position to see that they remained connected. Johnson set out to make the South a different kind of place. He sought to use massive infusions of federal money with strings attached to change its base from agriculture and poverty to research and technology. A side benefit, he hoped, was that creating more wealth would ease the South's racial tension. He thought that bringing white and black people together in the workplace would help with that, too. The NASA story demonstrates this plan in process.

Can one conclude, as Taylor said, that NASA was "outstandingly successful" in its equal employment efforts and its broader execution of federal antidiscrimination policies? There are many ways to answer that question. Hard numbers do exist. Census figures break down how many black and white engineers, computer specialists, scientific technicians, and the like there were in the communities in which NASA had a significant presence. Those numbers give one picture. In addition, there are the impressions of NASA's African American pioneers themselves. They give a different sense of what the space program did, what it did not do, and what it might have done. It is also possible to look at the impact NASA had on the communities as a whole. After all, if one believes that a rising tide lifts all boats, this metric can help shape an impression, too.

Several factors allowed NASA as an agency to promote civil rights. Policy enforcement, leadership by agency chiefs, NASA's public image as a modern and enlightened meritocracy, and the political-economic influence the agency held over contractors and within some host communities all contributed. Each factor allowed NASA, with varying degrees of success, to advance the cause of racial equality in the South.

Political and economic influences are basic tools of government agencies and critical industries within a community, but if the community is not centralized, neither will be fully effective. Only in Huntsville did conditions exist that allowed NASA to have a significant impact on local race relations. MSFC director Wernher von Braun and his German rocket men had lived in Huntsville since 1950. They were the core of the NASA workforce at MSFC. When von Braun spoke against segregation, he did so as a federal executive and, more importantly, as a respected neighbor with a long record of civic activities. No other NASA center director, regardless of competency, had a term of residency or community service similar to von Braun's in Alabama. Neither did any other host community have as effective a business coalition as AHAC, the Association of Huntsville Area Contractors. NASA and its contractors became Huntsville's second skin and protected the city from many of the worst segregationist excesses and from the wrath of George Wallace during the 1960s. It is difficult, however, to say they had a role in changing the racial makeup of the community's workforce, and there are even suggestions that, when it came to integration, the MSFC personnel office continued to stand in the way.

In a paper about NASA's civil rights activity in the years after the Moon landing, Kim McQuaid recounted the firing of Ruth Bates Harris, the assistant deputy director who dealt with equal opportunity issues at NASA headquarters in the early 1970s. The agency fired Harris for "submitting a private report to NASA Administrator James Fletcher stating that NASA's belated equal opportunity program was 'a near total failure.'" The report paints a particularly damning portrait of the hiring at MSFC. The center, it said, "appointed only one totally inexperienced employee rather than the three highly qualified specialists required." The report found that the center's EEO staff "has been continuously kept short of resources and under the control of insensitive middle management."[6]

Charlie Smoot, for his part, said that NASA in Alabama could have done much more to increase the number of African Americans it had working at Marshall. He said the constant whining from the agency about how it could not find quality blacks to hire was nonsense. "It was not easy," he said, "but you could find them if you wanted to. It depended on how bad you wanted to." When NASA said they could not get blacks to come to Alabama, he said, "they used that as reasons for not doing the right thing. They used it as an excuse. It gave them an out, a reason why they didn't do anything." The truth, he said, is that "they didn't do it because it was difficult." Smoot said he dealt with constant frustration over what he saw as a lack of willingness on the part of the MSFC personnel staff to accept the people he would send their way. He even had trouble placing his vaunted "Water Walkers."

"They would not take them," he said. "You'd bring them credentials and they would drag their feet."[7]

Nevertheless, Huntsville was a different kind of place than the rest of Alabama, a point that Governor George Wallace's segregationist rhetoric and confrontational style gave the city and its principal employer ample opportunity to make. Wallace was a lightning rod. His views allowed NASA as an agency to enter racial politics without fear of a presidential leash or serious retribution. The agency's economic and political presence also allowed Huntsville to resist Wallace, whereas other Alabama communities that lacked a "big brother" succumbed easily, perhaps willingly, to segregationist pressure. Finally, NASA's prestige and the powerful symbolism attached to the agency and its mission launched Huntsville onto national and international platforms. Consequently, civic leaders recognized the importance of public image to long-term economic growth and coordinated their plans with the agency to accomplish mutual goals.

When they got together in 2008 to discuss their experience as some of NASA's first African American employees, Richard Hall, Delano Hyder, and E. C. Smith all concurred in this assessment. NASA changed Huntsville, they said, but it did not change Alabama or the South. As Smith said, "It changed in Huntsville. Yes. How far did that spread? I don't know. I doubt very far." Huntsville, Hyder said, "was different in the beginning, but with the space program coming into Huntsville it even made it more special and different." His feeling (and the others chimed in to agree) was that the city's educated, multinational workforce made it unique.[8]

The Wallace situation did not repeat itself in other host states. Florida governor Farris Bryant promoted the state's ethnic diversity. Texas governor John Connally was loath to embarrass Lyndon Johnson. And finally, the Mississippi Test Facility (MTF) was far removed from Mississippi's political and social strife and had so few NASA employees that it was insulated from the problems of the day.

Public image and Space Age rhetoric sold the human space program to the American people, and to a degree, it sold desegregation as a de facto part of modern society to southern communities. The most successful sale was Brevard County, Florida. Here more than anywhere else in the South, Space Age imagery outweighed substance as an agent of social change. Regardless of NASA activity, the agency's very presence in Florida inspired reformers to invoke the agency's modernist image to advance the cause of racial equality.

NASA and contractor personnel at Cape Canaveral, unlike their Alabama counterparts, remained apart from local politics and social issues. One reason for this disengagement was the apolitical nature of engineers. Another reason was the term of residency—the Florida personnel viewed their time at Cape Canaveral as one more duty station in the course of their careers. The space workers were in Brevard County to engineer rocket launches, not re-engineer local society. Julius Montgomery's decision to pull out of Brevard Engineering College was one demonstration of that. The "old-line . . . rural, southern, here all their lives" culture described by James W. Button in *Blacks and Social Change* was another."[9]

Politics and money were not absolute determiners of racial moderation. NASA spent hundreds of millions of dollars in Brevard County but never intervened in local issues, as it did in Huntsville. The Cape Canaveral workforce was not concentrated in a single community, but spread throughout the county. These workers were more interested in roads and sewer lines than in political power and social reconstruction. Space imagery in other communities was seldom attached to the civil rights movement. At Huntsville, the imagery was more mainstream and served to reinforce the national effort to define the agency and its goals in heroic, even mythic, terms. Even the comments Webb and Wernher von Braun made in the mid-1960s related primarily to policy and economics rather than functioning as symbols of the Space Age.

What area of integration *did* NASA influence in Florida if not schools, housing, employment (including its own), and public accommodations? What did NASA contribute? The space program's in-migration of space-related workers and their families accelerated an already rapid population growth. The agency, its contractors, and the newcomers themselves increased tax and sales revenue and added to community prosperity. The community responded with improved roads, sewage systems, and greater spending on schools, though not always in black residential areas. It can be argued, however, that the general quality of life in the community improved.[10] In Titusville and Brevard County, the space program caused the growth that led to better community services. A black resident later said, "Racial changes came about because of changes in the community due to the Cape. You can't live in a Space Age and treat people in antiquated ways. We have better human relations now because there are more people in the community who care about all the people."[11] Clearly the Space Age and, by implication, program-related personnel were seen as agents of change. Another impact is that in 1969 Julius Montgomery became the first African American ever elected to the Melbourne City Council. When he retired in 1977, however, he was still the council's only black member, and, after he left, they did not have another for thirty-five years.[12]

NASA's role in the advancement of racial equality in Brevard County lacked the fire and thunder of its rocket launches. The agency was more facilitator than agitator. Brevard County slowly changed itself. NASA's arrival, replete with dollars and symbols, accelerated the process. Cities and towns that benefited from NASA dollars could no longer use low revenues to excuse race-biased services. The idea that outsiders changed places like Brevard County was a convenient conceit for all involved. On the one hand, traditionalists could blame "Yankees." On the other, reformers used the myth to shame the federal government into action; and most important for the agency, NASA's public image remained untarnished.

Houston was understandably proud of its space center and quickly engaged space rhetoric for its very own. But in "Space City," the rhetoric and hyperbole promoted business and industry—not civil rights and racial harmony. Houston had begun a slow process of desegregation before NASA arrived. And the city warmly embraced the Manned Spacecraft Center (MSC) and its contribution to the local economy. However, Houston's population was too large and its economy far too strong and diverse for NASA to dominate. Only the integration of Rice University can be directly linked to NASA's influence. Even if NASA officials wanted to challenge segregation in Texas and Houston, the political cost would have been enormous. NASA's keenest political supporters were Texans, some from the Houston area, who would not appreciate agency interference. Thus, political influence inhibited rather than fostered NASA intervention.

Other than integrating cocktail parties at the Petroleum Club, promoting limited training/educational opportunities, and the alteration of the Rice University charter, of what direct benefit was NASA to the average black worker and resident of Houston? NASA as a singular entity benefited blacks no more than it did whites. The agency was not a dominant player and thus could not leverage the local community on questions of race as it did at other southern installations. Even NASA's image as a champion of modernity and change—so popular in Florida—was little used in Houston. Furthermore, because so much of Houston had desegregated or was in the process of desegregating by the time of NASA's arrival, there seemed little for the agency to do. What NASA could do and did do was to enforce its own policies and oblige those who wished to do business with the agency to adopt desegregated, if not integrated, policies of their own. Finally, the agency required the support of Texans in Congress and the White House, support that might waver if the MSC became active in racial politics.

No two NASA site locations were more different than rural Mississippi and metropolitan Houston. Yet in both places, the space agency's prestige and money failed to radically alter long-standing racial attitudes and policies. Despite the NASA payroll, Pearl River and Hancock Counties remained too isolated and too small in native population for the agency to change social tradition. On the other hand, Houston's large population and its powerful economic base, so attractive to the selection committee, granted the Bayou City immunity from a Space Age social agenda. Different circumstances produced the same result—a Space Age business as usual.

NASA always said that its greatest problem with equal employment was not the hiring of African Americans but finding African Americans to hire. The agency's black pioneers suggest that NASA was not looking in the right places, but hard numbers suggest it was going to be an effort no matter what. Regardless of who was right, NASA's employment efforts were paltry. In 1961, the agency employed 18,953, of whom 3 percent were black. That percentage remained stable as the decade progressed. In 1964, 2.9 percent of NASA's employees were black. In 1965, 3.2 percent were black. Agency downsizing reduced employment in 1967, but in that year, still 2.9 percent were black. By the time of Apollo 11 in 1969, 3.1 percent of the NASA workforce was African American.[13]

None of NASA's African American pioneers interviewed for this book, with the exception perhaps of Charles Smoot, had any knowledge of the greater national picture when it came to the pool of African Americans from which NASA could draw. They were going with their gut. They all had skills and they went to school with people who had skills. They merely assumed there must be many more. Census data suggests otherwise.

The near complete absence of black engineers and technicians in some southern states made an integrated workforce an almost impossible goal, even though NASA did establish local training programs within host communities. Census data for NASA's southern host communities from 1960 shows what the agency faced. Data from the 1970 census shows how well (or poorly) Johnson did in his plan of creating a new category of jobs for blacks to hold. The following highlights demonstrate how significant the hurdle was.[14]

When the government conducted the 1960 census, there were five African American electrical and electronic technicians in the entire state of Florida. The state had 3,734 of them overall. The state had no African American aeronautical engineers. Not one. It had thirteen black electrical engineers and nine black mechanical engineers.[15] The situation had improved ten years later, though not by much. First, there were new categories of jobs—computer specialist, astronautical engineer, scientific technician—created by the space program. There were African Americans who held jobs in all these fields. The state had 4,721 people working as computer specialists, 88 of whom were black. There were 2,136 aeronautical and astronautical engineers, 24 of whom were black. Nearly 21,000 new engineering and scientific technician jobs were created in Florida during the Space Age. Of those, blacks held 488. The number of African American electrical engineers had gone from thirteen to thirty-nine. The number of African American mechanical engineers went from nine to thirty-nine.

As memo after memo from NASA and the Huntsville contractors lamented, black professionals and technicians in Alabama were almost unknown in 1960. The state had four black aeronautical engineers, four black electrical and electronic technicians, and no African American designers or draftsmen. Those numbers are not for Huntsville, but for the entire state of Alabama.[16]

NASA remained concerned with its local influence and race relations within Alabama throughout the remainder of the decade. However, by 1966 changing federal priorities and agency concern for continued Project Apollo funding overshadowed other considerations. Webb continued his criticism of Alabama's image, raising the ire of citizens and politicians to the point where Alabama congressman

Robert Jones asked the president, "Why is it necessary to alienate our people in this manner? Such divisive statements seem so unnecessary and they serve no useful purpose."[17] Despite Webb's criticisms, however, the employment situation at NASA in Alabama appears to have remained static for several years into the 1970s as well.

By 1970, black professionals and technicians remained an insignificant part of the state's workforce. Out of 2,300 computer scientists, only 33 were black. There were six African American aeronautical and astronautical engineers in the state by 1970, only 252 engineering and scientific technicians, and only 25 electrical and electronic technicians. There were thousands of whites working in all of those jobs.

As in the other locations, the low number of black professionals and skilled technicians employed in pre-NASA Houston represented a shallow labor pool from which to draw. Houston had no African American aeronautical engineers in 1960 and no black electrical or electronic technicians. The census showed there were nine African American draftsmen and eight electrical engineers. The absolute number of black professional and skilled technicians increased between 1960 and 1970, but federal employment of these workers was abysmal—abysmal not only in relation to private industry but also when compared to Mississippi. The federal government in Houston employed no African American civil engineers, no African American mechanical engineers, no African American life and physical scientists, and no African American chemists. The federal government employed six black computer specialists, five astronautical engineers, and only nine engineering and scientific technicians, while employing thousands of whites in these job categories.[18]

Part of this may have had to do with the very location of the MSC. As Otis King said, "NASA is located at Clear Lake, which is at least thirty miles from downtown Houston." African Americans were less likely than whites to own their own cars in the early 1960s,[19] but aside from that, as Otis King said, Clear Lake was in the middle of "Ku Klux Klan territory." There were African Americans in Galveston, he said, and "some blacks in Texas City, but not much in between and so you had a stretch of about fifty miles where there were blacks in Houston and then blacks again in Galveston, but no blacks living in between and NASA of course was right in the middle of that in space."[20]

To understand the absence of blacks at NASA's facility in Mississippi, one must accept two facts: first, the recruitment of black professionals for work in rural Mississippi in the 1960s was difficult at best, and second, black professionals and technicians were a rarity in 1960 Mississippi.[21] In 1960, the state had no black chemists, no black draftsmen, no black electrical or electronic technicians, no black electrical engineers, and only four black aeronautical engineers.[22]

Sadly, NASA did not significantly alter attitudes in Mississippi, either. The hardened positions Peter Dodd found in 1963 continued throughout the decade. Despite published reports of official optimism that the MTF would lead to changes in social customs in southern Mississippi, no evidence suggests that a statewide or regional change in racial attitudes occurred because of the space program. NASA contracts provided the federal government with additional but unused leverage in its dealings with Mississippi. If the agency had an impact on Mississippi racism, it did so indirectly and in a manner quite different from Alabama's clash of personalities and Florida's symbolism. By 1970, the number of employed black engineers and technicians in Mississippi had increased from 1960 but not by much. There were 170 chemists, 16 draftsmen, 4 electrical or electronic technicians, and 16 electrical and electronic engineers. That is better, but nothing to brag about. The state still had no aeronautical and astronautical engineers.[23] The increase in statewide totals and the inclusion of blacks in all save one category cannot be solely attributed to the economic growth of a racially enlightened Mississippi. At best, the 1970 census numbers were a hollow victory for reformers. The number of black professionals compared to whites remained pathetic. The one area where the MTF could have truly changed Mississippi was through employment. It failed. By 1970, the federal government employed only 95 of the 364 black scientists, engineers, and technicians in the state. Even if NASA or its contractors employed all 364, the number would represent only 12 percent of the total MTF employment in 1968. Change was slow in Mississippi.

But the overall scorecard is not all bleak. There was massive hiring during the space program, and it was not limited to the hiring of scientists and engineers. Herbert R. Northrup, of the Wharton School of Finance, tracked the hiring of African Americans in aerospace throughout the period of NASA's greatest growth. His writing shows one of the rays of light in the story of NASA's impact on the African American community. Writing at the end of the decade in the *Monthly Labor Review*, Northrup conceded, "Aerospace companies have scoured the country looking for professional and technical employees. But Negroes have traditionally not been oriented to engineering as a profession." He added, "The few who are found in the aerospace industry are a sizable percentage of those available" and opined that many of the employed black engineers who graduated from segregated schools did not have a background conducive to corporate mobility and worked where they did, in part, due to governmental pressure.[24]

Nevertheless, while the federal government may not have been able to create new kinds of jobs in the South that African Americans could hold, as Northrop stated in *The Negro in the Aerospace Industry*, technology jobs are not the only jobs out there. In addition to scientists and engineers, Northrop wrote, the aerospace industry required "huge clerical, accounting, finance, personnel, systems and computer and data processing staffs to operate and to control its manufacturing, research and testing facilities." This need had "a distinct impact on the industry's capacity to employ Negroes," he said, "since Negroes are disproportionately unrepresented among these groups in our society.[25]

As a result, one can find at least part of the change Johnson and others were looking for. The federal government's presence created jobs, and many of those jobs went to African Americans. Northrup discovered that in the Southeast in 1968, major aerospace companies had opened facilities and "partially because they have practiced equal employment, and partially under federal government prodding, they have changed employment practices of the region in a major manner." Even though African Americans were not working in most of the new high-tech jobs, these aerospace companies "have done much more in this regard than many other industries, for example automobiles, which have tended much more to maintain the status quo."[26]

Could NASA have done more for blacks more quickly than it did during the 1960s? The agency could only advance civil rights or any other policy beyond its charter as quickly and as strongly as the president, Congress, and the agency's administrator would allow. NASA followed when and where the Kennedy and Johnson administrations led federal intervention into the racial equality campaign. The space agency followed not only as one of many federal agencies but as an agency that enjoyed and wished to maintain almost unqualified presidential and southern congressional support. Independent action by NASA in the politically sensitive area of civil rights, without the umbrella of presidential protection and later the Civil Rights Act of 1964, would have—in all probability—doomed the agency. So yes, NASA could have done more to advance racial equality more quickly *if* the federal government as a whole had been more willing or able to address the issue. Despite the efforts of Presidents Kennedy and Johnson and the NASA administration, race relations progressed less quickly than space technology and the attainment of space goals.

What is NASA's excuse after the tumult of the 1960s subsided? Well, NASA's post-1960s employment record is much more difficult to justify. Looking over NASA's record in the 1970s, 1980s, 1990s, and early twenty-first century, James Jennings found that "NASA hasn't made the progress that it should have as far as minorities and them moving up into positions. In fact," he said, "if you look at the percentages, it's not very much different today than it was probably twenty years ago."[27]

NASA had 5.6 percent minority employees in 1973, versus a government average of 20 percent. At that time, Deputy Administrator George Low is said to have declared, "Equal Opportunity is a sham in NASA."[28] By 1983, the number of African Americans at the agency had doubled, from 6 percent of NASA's labor force to 12 percent. In the science, engineering, and technical half of NASA, the increase was from 3.9 percent to 8.3 percent. By 1991, about 4 percent of all NASA science and engineering jobs were held by African Americans.[29]

This all addresses the question of what impact civil rights had on the space program. But turning the question around is important, too. What impact did the space program have on civil rights? To answer this, one must turn again to the plan by southern New Deal politicians to use government spending to change the South. The space program helped end the protracted brain drain and culture drain of African Americans that Americans have come to know as the Great Migration. Beginning in 1916, 6 million African Americans, fleeing treatment that "doesn't warrant staying," left the Jim Crow South.[30] When the Great Migration began, 90 percent of all African Americans were southerners. By the time it was over, nearly half were living somewhere else. The Great Migration ended in the 1970s.[31] The space program was one of the reasons why.

The Great Migration empowered millions of African Americans by giving them the rights they were denied in the places of their birth. It also enriched the nation with the contributions of the children of those first pioneers. As Isabel Wilkerson, author of the epic chronicle of the Great Migration *The Warmth of Other Suns*, has said, this massive movement of people "changed our culture, it changed the music that we listened to, it changed literature, it changed politics."[32] But, as inevitably happens whenever communities are uprooted, the movement wiped out important elements of the culture that had existed in the South before.

Morgan Watson saw these reluctant escapees while he was growing up. They would "come back to the South on visits," he said. He heard and experienced their sadness, the loss of leaving the place they understood. Watson made and kept many vows throughout his life to end what was wrong with the place where he grew up. He saw going to work at NASA as yet one more. Watson said of himself and the other African American NASA co-ops, "We were tired—all of us—of the outmigration."[33]

Before the Space Age, there really was no alternative. As James Jennings remembered, when he was growing up, "the mentality was: I can go to the North and get a good job and live in a nice place and— you know, everything will be great." He talked about a man he knew. "He was two or three grades ahead of me, name was Stanley Kennedy." Kennedy, he said, "went to Tuskegee and he was a brilliant guy. I think he made one B during his whole, entire education career." When his friend graduated, Jennings said, he "went to work for IBM up in Rochester, New York. My thinking and probably the thinking of many other people at that time was, If a guy this brilliant can't get a job here in the South, I certainly can't. So I need to be looking to go somewhere in the North also."[34]

Like many of the 6 million who left, Jennings and Watson loved the South. They considered themselves people of the South, and Watson said that although he "got very good job offers in other parts of the country," because of the changes brought by NASA and the rest of the Second Reconstruction, "those of us that wanted to remain in the South; it gave us an opportunity to do so."[35]

Isabel Wilkerson has said that the African Americans who left the South gave a hand to those who stayed in helping them change the region's conditions. While "six million people actually voted with their feet" and became a force that pushed northern and western Democrats to support civil rights, "those who actually stayed [in the South] marched and protested for things to change." That is what transformed the South. These two groups together, she said, "were actually able to do what the Emancipation Proclamation on the day that it was signed could not do."[36]

But the African American pioneers of the space program represent a third group. The people who went to work for NASA and the space program in the South did not march. They did not protest.

Morgan Watson and George Bourda sat out the sit-ins in Baton Rouge. Julius Montgomery capitulated to segregation at Brevard Engineering College (BEC). Working through agency channels, Clyde Foster got NASA to create a separate but equal advanced training program for blacks in 1970. After Martin Luther King Jr. was assassinated and Washington, DC, burned, not only did astrophysicist George Carruthers not get involved; he was so distracted with his scientific discoveries that he did not even realize the riots occurred.[37] No, the space program's African American pioneers were not marchers. As Delano Hyder said, people in Huntsville had "a different attitude" when it came to protesting. As a result, as Richard Hall said, protesting was something that "just doesn't become a part of you."[38]

There are a number of ways to look at this dichotomy. In the case of Foster and Montgomery, they were not conciliatory; they were strategic. Montgomery stayed away for a year. Once BEC got on its feet, he enrolled. Now the school has an award named after him. If he had called in SNCC or the NAACP and caused a fuss, none of that could have happened. Clyde Foster appealed in person to Eberhard Rees, the Marshall director who replaced von Braun, and got him to bring African American professors from all over the South to Alabama A&M to teach black NASA employees. The result was that Hall, Hyder, Foster, and others all got promotions and raises. Foster could have organized a walkout; he could have marched with a placard back and forth in front of the gates at MSFC. He saw the benefit—the greater benefit to all—of holding back.

The people who push through the door are not always the people who march. Sometimes they are the ones who show up every day, work hard, do their jobs, and impress on everyone around them that they can handle the load. "We had heard Martin Luther King say, 'When you start off a race behind, you have to run faster than everybody else,'" Morgan Watson said.[39] He and the other African American co-ops knew—as Clyde Foster and Julius Montgomery and all of the others knew—that their success was not just personal; nor were their failures. If they messed up, the repercussions were magnified. It was not "I messed up." It was "those people can't be trusted." They had to be the very best and the very best at all times. As Watson said, they knew that they could not fail.

The national media—especially the black media—took note. The black press had always been the biggest cheerleader in African American communities, confirming for a downtrodden people that what they saw and feared was reality was, in fact only a manifestation of their repression; that they could do anything if only someone would give them the chance. The coming of the space program handed the black press two new tools of racial uplift. The cutting-edge jobs created by NASA and its contractors allowed for the creation of a whole new crop of heroes. The papers held up Ed Dwight, Frank Crossley, Clyde Foster, George Carruthers, Morgan Watson, and George Bourda as icons of intelligence, worthiness, and ability. And especially in the case of Dwight, the black press was able to use the futuristic imagery of the Space Age and turn it back on American society to force the question of whether the nation truly was launching into the future or whether it was still stuck in the past. As for the mainstream white media, when the *New York Times* called Frank Williams "a social pioneer" and "a symbol," it was giving notice to white America that its black fellow citizens possessed the requisite level of intelligence to compete at the highest level of American science and technology.[40] As someone who passed through the center of all this change, Morgan Watson saw the impact both short term and long. Today, he said, "no matter where you go in the South, you see a great number of black professionals."[41] That was not the case before Kennedy and Johnson began creating jobs—including brand new types of jobs, which African Americans could hold.

Morgan Watson had the opportunity to drive this point home in a particularly meaningful way in an event at the National Air and Space Museum on February 20, 2010. Of the almost fourteen thousand people who streamed in to the museum that day, many were black.

February 20 was African American Family Day, a special event for patrons to learn things they had probably never heard before about African Americans in aerospace history. Highlights included a talk with surviving members of the Tuskegee Airmen and a panel featuring the first all African American, all-female air crew. There was one more very important panel, too.

NASA finally opened the ranks of its most exclusive club, the astronaut corps, to African Americans in June of 1967. Air force major Robert H. Lawrence, the first black astronaut, died in a plane crash five months later. But the creation of the shuttle program finally allowed NASA to send African Americans into space. Dr. Guion Stewart "Guy" Bluford Jr. became the first African American to do so in 1983, and Dr. Mae Jemison, the daughter of a roofer who worked on the construction of Huntsville's Redstone Arsenal, became the first African American woman in space in 1992.[42]

The National Air and Space Museum panel featured astronaut Mae Jemison, along with astronaut Leland Melvin, just back from a trip to the International Space Station. Sitting on the dais with them were two special guests: Morgan Watson, then in his mid-sixties, and Julius Montgomery, still amazingly spry at nearly ninety. The museum designed family days to teach younger patrons something new, but, as it would turn out, this family day would be a learning experience for the younger panelists—that is, the astronauts—too. Today they would hear first-hand stories of just what it took to kick in the doors that allowed them to eventually reach the stars.

Watson talked about picking cotton in ninth grade, how the buses would ride by on the roads, filled with white kids shouting taunts. And Montgomery talked about being the first African American at the Cape way back in the '50s. What it felt like to realize that "I was always the Only One. Where I worked and wherever I went, I was the only one." And, as he always did in settings like this, he told the story of his first day on the job at RCA, knowing there was probably no place worse for African Americans in Florida than the area around Cape Canaveral, that if you were black, you had a greater chance of being lynched there than anywhere else in the South. He talked of walking up to each man, and having them turn their backs; how no one would shake his hand and how his "great white bastard" comment got everyone to laugh and finally broke the ice. The two astronauts listened with rapt attention.

Melvin shared a funny story of a white man who ignored him in an elevator at NASA only to learn later in a meeting that he had snubbed an astronaut. He and Jemison reminisced about how, as blacks, it was frightening to drive the back roads around the Cape.

Later on in the Green Room, as people snapped pictures of Mae Jemison with the airmen (it was tough to tell who was more honored to meet whom), Julius Montgomery sidled over to Leland Melvin and looked up (the astronaut, a former wide receiver for the Detroit Lions, towered over the older man) with wide-eyed awe. "I'll tell you," Montgomery said, "You astronauts; you're the bravest people I ever met." Leland Melvin returned the look and his grin broke into a wide, beaming smile. "No, sir," he said. "I heard your story out there. You are the bravest person I ever met." And Julius Montgomery laughed and Leland Melvin laughed, and they shook hands.[43]

Here, in essence, was the interplay between the space program and civil rights. Bounded in a nutshell was Julius Montgomery, whose bravery opened the door to let Morgan Watson's bravery break the walls down and allow Mae Jemison and then Leland Melvin to count themselves rulers of infinite space. This is the very progression that Lyndon Johnson and Hobart Taylor had envisioned and the one they pressed on James Webb and Wernher von Braun to create.

Did it do inestimable good? Ask those who benefited—ask Mae Jemison and Leland Melvin. Ask former astronaut Charles Bolden, who became NASA's first African American administrator in 2009. Was it perfect? Of course not. Overall, Charles Smoot said, "there was some good done by NASA but they were no better than Martin Marietta or Lockheed or any of the others." There were changes made by NASA, he said, but "they weren't the only one," and not only that, "they did it screamin' and hollerin'." In the end, he said, "I'm trying to be fair. At that time, it was not easy for them. NASA did a lot of good. They could have done better."[44] Those who benefited from NASA's action, however, took the opportunity given to them and ran with it as far as they could go.

America had to pursue the space program, President Kennedy said, because there was "new knowledge to be gained and new rights to be won and used for the progress of all people."[45] While he was not talking about American race relations when he said that, an accident of timing and coincidence ensured that the space program would help win rights and create progress for African American people in ways the president could not have imagined. In doing so, the space program would help white Americans gain new knowledge about their black fellow citizens and their abilities. In the end, NASA's story with civil rights is just one agency's story. The plan put in place by Kennedy and Johnson caused the federal government to hire African American men and women throughout the South during the turmoil of the 1960s. There are doubtless stories of other Water Walkers out there; people who also helped break the walls down and change the perception of black people in the South.

Appendix

Relevant Census Numbers on Employed Professional and Skilled Labor for NASA Host States

Table 1. Employed Professional and Skilled Labor in Alabama and Huntsville, 1960

Occupation	State	Black	Huntsville total
Chemist	967	9	Not available
Designer or draftsman	2,084	0	Not available
Aeronautical engineer	523	4	326
Civil engineer	2,680	13	121
Electrical engineer	2,150	13	643
Mechanical engineer	1,717	9	637
Other engineer	3,644	13	252 (industrial)
Electrical or electronic technician	904	4	Not available

Source: U.S. Department of Commerce, *1960 Census of the Population, Vol. 1: Characteristics of the Population, Pt. 2: Alabama* (Washington, DC: Government Printing Office, 1963), 2-373, 379, tables 121 and 122.

Table 2. Employed Professional and Skilled Labor in Alabama, 1970

Occupation	State	Black	Private employer (black/total)	Federal employer (black/total)
Computer specialist	2,300	33	17/1,654	11/514
Engineer	17,073	191	82/10,370	57/4,908
Aeronautical or astronautical engineer	1,051	6	0/465	6/578
Civil engineer	3,253	50	5/1,347	5/580
Electrical or electronic engineer	3,956	30	10/2,610	20/1,253
Mechanical engineer	2,517	26	18/1,754	3/702
Other engineer	6,296	85	49/4,194	23/1,785
Life and physical sciences	1,798	109	99/1,071	0/140
Chemist	1,050	77	73/864	0/140
Engineering or scientific technician	9,223	252	189/6,323	16/1,418
Draftsman or surveyors	3,941	108	85/2,942	6/200
Electrical or electronic technician	1,687	25	19/915	6/689

Source: U.S. Department of Commerce, *1970 Census of the Population, Vol. 1: Characteristics of the Population, Pt. 2: Alabama* (Washington, DC: Government Printing Office, 1973), 2-623, 626, table 173.

Table 3. Employed Professional and Skilled Labor in Florida, 1960

Occupation	State	Black
Chemist	917	12
Designer or draftsman	4,510	13
Aeronautical engineer	657	0
Civil engineer	3,736	30
Electrical engineer	4,970	13
Mechanical engineer	2,336	9
Other engineer	4,281	9
Electrical or electronic technician	3,734	5

Source: U.S. Department of Commerce, *1960 Census of the Population, Vol. 1: Characteristics of the Population, Pt. 11: Florida* (Washington, DC: Government Printing Office, 1963), 11-469, table 122.

Table 4. Employed Professional and Skilled Labor in Florida, 1970

Occupation	State	Black	Private employer (black/total)	Federal employer (black/total)
Computer specialist	4,721	88	60/3,815	13/323
Aeronautical or astronautical engineer	2,136	24	19/1,828	5/283
Civil engineer	4,422	47	23/2,455	6/440
Electrical or electronic engineer	9,093	39	34/7,931	5/831
Mechanical engineer	3,957	39	39/3,439	0/406
Other engineer	11,084	75	52/8,076	15/1,276
Life and physical sciences	3,166	47	21/1,550	4/796
Chemists	1,333	9	4/985	0/115
Engineering or scientific technician	20,764	488	354/15,868	40/1,793
Draftsman or surveyors	9,264	171	115/7,043	4/286
Electrical or electronic technician	6,551	126	111/5,177	11/1,012

Source: U.S. Department of Commerce, *1970 Census of the Population, Vol. 1: Characteristics of the Population, Pt. 11: Florida, Section 2* (Washington, DC: Government Printing Office, 1973), 11-1679, 11-1683, table 173.

Table 5. Employed Professional and Skilled Labor in Mississippi, 1960

Occupation	State	Black total
Chemist	180	0
Designer or draftsman	781	0
Aeronautical engineer	9	4
Civil engineer	1,299	4
Electrical engineer	468	0
Mechanical engineer	359	3
Other engineer	880	0
Electrical or electronic technician	155	0

Source: U.S. Department of Commerce, *1960 Census of the Population, Vol. 1: Characteristics of the Population, Pt. 26: Mississippi* (Washington, DC: Government Printing Office, 1963), 26-333, table 122.

Table 6. Employed Professional and Skilled Labor in Mississippi, 1970

Occupation	State	Black	Private employer (black/total)	Federal employer (black/total)
Computer specialist	540	4	0/395	0/59
Engineer	106	0	0/79	0/22
Aeronautical or astronautical engineer	1,585	39	0/381	23/634
Civil engineer	1,075	16	10/860	0/167
Electrical or electronic engineer	781	5	5/642	0/114
Mechanical engineer	1,988	14	11/1,654	3/200
Other engineer	1,114	28	5/491	18/286
Life and physical sciences	333	12	0/249	12/42
Chemist	4,445	170	103/2,882	30/706
Engineering or scientific technician	2,210	56	35/1,516	5/267
Draftsman or surveyor	504	16	12/288	4/162
Electrical or electronic technician	540	4	0/395	0/59

Source: U.S. Department of Commerce, *1970 Census of the Population, Vol. I: Characteristics of the Population, Pt. 26: Mississippi* (Washington, DC: Government Printing Office, 1973), 26-532, 535, table 173.

Table 7. Employed Professional and Skilled Labor in Houston, 1960

Occupation	Total	White	Black
Chemist	872	872	0
Designer or draftsman	3,037	3,028	9
Aeronautical engineer	35	35	0
Civil engineer	1,587	1,579	8
Electrical engineer	942	934	8
Mechanical engineer	1,527	1,523	4
Other engineer	4,465	4,465	0
Electrical or electronic technician	378	378	0

Source: Department of Commerce, *1960 Census of the Population, Vol. 1: Characteristics of the Population, Pt. 45: Texas* (Washington, DC: Government Printing Office, 1963), 45-884, table 122.

Table 8. Employed Professional and Skilled Labor in Houston, 1970

Occupation	Total	Black	Private employer (black/total)	Federal employer (black/total)
Computer specialist	3,503	83	73/3,252	6/158
Aeronautical or astronautical engineer	821	5	0/430	5/391
Civil engineer	2,868	21	21/2,412	0/37
Electrical or electronic engineer	3,283	56	45/2,891	11/318
Mechanical engineer	2,676	14	14/2,469	0/160
Other engineer	9,589	65	60/8,720	0/438
Life and physical sciences	4,799	80	80/4,168	0/153
Chemist	1,689	66	66/1,617	0/10
Engineering and scientific technician	11,447	427	388/10,401	9/299
Draftsman and surveyor	5,523	128	102/5,060	5/21
Electrical or electronic technician	1,733	56	52/1,480	4/183

Source: Department of Commerce, *1970 Census of the Population, Vol. 1: Characteristics of the Population, Pt. 45: Texas, Section 2* (Washington, DC: Government Printing Office, 1973), 45-1679, 45-1683, table 173.

Notes

Introduction

1. James L. Hicks, "Negro Math Expert Helped Launch Spaceman," *New York Amsterdam News*, May 13, 1961, 1, col. 1.

2. Claude Sitton, "Segregationists Fight New 'Lost Cause': Governor Wallace's Action in Closing Alabama's Schools Points up the Last-Ditch Resistance," *New York Times*, September 8, 1963, E5.

3. Konrad Dannenberg, conversation with Richard Paul, December 19, 2006.

4. "Astronaut Trainee Itching to Go into Orbit for US," *Los Angeles Sentinel*, June 6, 1963, 1, col. 6; "Negro College Youth to Boost First Moon Goer into Orbit?" *Chicago Defender*, March 14, 1964, 1, col. 5; "Space City Faces School Segregation Showdown," *Daily Defender*, March 28, 1963, 1, col. 11; "Her Science Paper Is Key to Man in Orbit," *New York Amsterdam News*, May 13, 1961, 1, col. 1.

5. Morgan Watson, conversation with Richard Paul, January 27, 2009.

6. James W. Button, *Blacks and Social Change: Impact of the Civil Rights Movement in Southern Communities* (Princeton, NJ: Princeton University Press, 1989), 3.

7. Johnson identified the agencies with the most contracts as "the Department of Defense, GSA, the Post Office Department, Veterans' Administration, the Space Agency." He also mentioned the Atomic Energy Commission. President's Committee on Equal Employment Opportunity, July 18, 1963, LBJ Library (hereafter LBJL), Austin, TX, 33, 38.

8. Hugh Davis Graham suggests Kennedy defaulted to PCEEO as a solution because he thought that "he could not get any significant civil rights legislation through Congress, and . . . therefore was unwilling to even try, lest he roil the Congress and threaten his higher priorities." Hugh Davis Graham, *The Civil Rights Era: Origins and Development of National Policy, 1960–1972* (New York: Oxford University Press, 1990), 28; Bruce J. Schulman, conversation with Richard Paul, December 2, 2008 (Schulman is the author of *From Cotton Belt to Sunbelt: Federal Policy, Economic Development, and the Transformation of the South, 1938–1980* [Durham, NC: Duke University Press Books, 1994]); Theodore J. Lowi, *The End of Liberalism: The Second Republic of the United States*, 2nd ed. (New York: W. W. Norton, 1979), 207.

9. Schulman, December 2, 2008; Gavin Wright, *Old South, New South: Revolutions in the Southern Economy since the Civil War* (New York: Basic Books), 199; Nancy MacLean, "From the Benighted South to the Sunbelt: The South in the Twentieth Century," in *Perspec-

tives on Modern America: Making Sense of the Twentieth Century, ed. Harvard Sitkoff (New York: Oxford University Press, 2001), 204; transcript, Robert S. McNamara Oral History, Special Interview I, March 26, 1993, by Robert Dallek, electronic copy, LBJL, 4.

10. James Jennings, conversation with Richard Paul, November 20, 2008.

11. Roger D. Launius, "Managing the Unmanageable: Apollo, Space Age Management and American Social Problems," *Space Policy* 24 (2008): 158.

12. David H. Onkst has focused on race relations in the late 1950s at the Grumman facility in Long Island, New York, while Kim McQuaid's work on racism at NASA focuses on the 1970s and 1980s.

13. Glen Asner, "Space History from the Bottom Up: Using Social History to Interpret the Societal Impact of Spaceflight," in *Societal Impact of Spaceflight*, ed. Steven J. Dick and Roger D. Launius (Washington, DC: NASA History Division, 2007), 387.

14. President's Committee on Equal Employment Opportunity, July 18, 1963, LBJL, 43.

15. George C. Wallace, "The 1963 Inaugural Address" (speech), January 14, 1963, Alabama Department of Archives and History, p. 2, http://www.archives.state.al.us/govs_list/inauguralspeech.html.

16. Wernher von Braun, "Huntsville in the Space Age" (speech, annual banquet), Huntsville/Madison County Chamber of Commerce, Huntsville, AL, 1964, 15.

17. Robert Cohen and David J. Snyder, eds., *Rebellion in Black and White: Southern Student Activism in the 1960s* (Baltimore: Johns Hopkins University Press, 2013), 3.

18. Asner, "Space History from the Bottom Up," 399.

19. Letter, James E. Webb to Wernher von Braun, June 24, 1963, Minority Groups (1961–1993), file 008983, NASA Headquarters Historical Reference Collection, Washington, DC; John W. Finney, "NASA Is Training Negroes for Jobs," *New York Times*, May 31, 1964, 54.

20. Theodis Ray, conversation with Richard Paul, February 16, 2009.

Chapter 1

1. Guy Bluford became the first African American in space in 1983 and Mae Jemison became the first African American woman in space in 1992.

2. "Kill Six in Florida; Burn Negro Houses: Search for Escaped Negro Convict Leads to Race Riot, in Which Two White Men Die," *New York Times*, January 6, 1923, 1.

3. Charles Payne, "You Duh Man! African Americans in the Twentieth Century," in *Perspectives on Modern America: Making Sense of the Twentieth Century*, ed. Harvard Sitkoff (New York: Oxford University Press, 2001), 178.

4. MacLean, "From the Benighted South," 208.

5. Ibid.

6. Ben Green, *Before His Time: The Untold Story of Harry T. Moore, America's First Civil Rights Martyr* (Gainesville: University Press of Florida, 1999), 36.

7. Theodore L. Reller, "The School and Child Welfare," *Annals of the American Academy of Political and Social Science* 212 (November 1940): 52.

8. Green, *Before His Time*, 21.

9. Jerrold M. Packard, *American Nightmare: The History of Jim Crow* (New York: St. Martin's Press, 2002), 171.

10. Roberta Senechal de la Roche, "The Sociogenesis of Lynching," in *Under Sentence*

of Death: Lynching in the South, ed. William Fitzhugh Brundage (Chapel Hill: University of North Carolina Press, 1997), 56.

11. MacLean, "From the Benighted South," 209.

12. Ben Green, quoted in "Race and the Space Race," Public Radio Exchange, http://www .prx.org/pieces/41113-race-and-the-space-race.

13. Green, *Before His Time*, 31.

14. William Fitzhugh Brundage, ed., *Under Sentence of Death: Lynching in the South* (Chapel Hill: University of North Carolina Press, 1997), 4.

15. David Colburn, conversation with Richard Paul, December 26, 2006. Colburn is the author of *From Yellow Dog Democrats to Red State Republicans: Florida and Its Politics since 1940* (Gainesville: University Press of Florida, 2007).

16. We asked all of the retired African American NASA employees interviewed for this book a variant of the question "When you were young, what kind of job could a black man aspire to?" The complete list of jobs gathered was teacher, janitor, cook, concrete worker, railroad man, foundry worker, post office worker, doctor, and lawyer. Bearing out this perception are statistics from 1960 showing that the entire state of Florida had five African American electrical or electronic technicians (Alabama had four), thirteen African American electrical engineers (Alabama also had thirteen), and zero aeronautical engineers (Alabama had four). U.S. Department of Commerce, *1960 Census of the Population, Vol. 1: Characteristics of the Population, Pt. 2: Alabama* (Washington, DC: Government Printing Office, 1963), 2-373, 379, tables 121 and 122; U.S. Department of Commerce, *1960 Census of the Population, Vol. 1: Characteristics of the Population, Pt. 11: Florida* (Washington, DC: Government Printing Office, 1963), 11-469, table 122.

17. NASA retiree Clyde Foster, in talking about Alabama A&M College's lack of willingness to participate in NASA-sponsored programs to promote engineering, said the school would only train engineers to "build some damn roads" because road construction was the principal job the school administration saw as being open to blacks with engineering degrees.

18. Julius Montgomery, conversation with Richard Paul, February 6, 2008.

19. Hamilton Bims, "Rocket Age Comes to Tiny Triana," *Ebony*, March 1965, 106; Lloyd Leigh, "Negroes Vital in Space Program," *Chicago Daily Defender*, November 30, 1965, 7.

20. In "Affirmative-Action: Past, Present, and Future," Peter H. Schuck said Congress authorized the National Labor Relations Board to redress an unfair labor practice by offering the offending party to "cease and desist from such unfair labor practice, and to take such affirmative action . . . as will effectuate the policies of this act." Peter H. Schuck, "Affirmative-Action: Past, Present, and Future," *Yale Law and Policy Review* 20, no. 1 (2002): 46.

21. Douglas Helms, "Eroding the Color Line: The Soil Conservation Service and the Civil Rights Act of 1964," *Agricultural History* 65, no. 2 (Spring 1991): 37, 39.

22. Graham, *Civil Rights Era*, 10.

23. Ibid.

24. David Hamilton Golland, *Constructing Affirmative Action: The Struggle for Equal Employment Opportunity* (Lexington: University Press of Kentucky, 2011), 40–41.

25. Merl E. Reed, *Seedtime for the Modern Civil Rights Movement: The President's Committee on Fair Employment Practice, 1941–1946* (Baton Rouge and London: Louisiana State University Press, 1991), 349.

26. Charles W. Eagles, "Review of Seedtime for the Modern Civil Rights Movement: The

President's Committee on Fair Employment Practice, 1941–1946, by Merl E. Reed," *American Historical Review* 101, no. 4 (October 1996): 1299.

27. Montgomery, February 6, 2008.

28. Payne, "You Duh Man!" 189.

29. Isabel Wilkerson, *The Warmth of Other Suns* (New York: Random House, 2010), 320.

30. Ibid., 323.

31. Larry J. Griffin, Paula Clark, and Joanne C. Sandberg, "Narrative and Event: Lynching and Historical Sociology," in *Under Sentence of Death: Lynching in the South*, ed. William Fitzhugh Brundage (Chapel Hill: University of North Carolina Press, 1997), 24–25.

32. Wilkerson, *Warmth*, 320–321.

33. William Gary, conversation with Richard Paul, December 21, 2006.

34. Paul Ortiz, conversation with Richard Paul, January 10, 2007 (Ortiz is the author of *Emancipation Betrayed: The Hidden History of Black Organizing and White Violence in Florida from Reconstruction to the Bloody Election of 1920* [Berkeley: University of California Press, 2006]); Green, *Before His Time*, 72, 29, 117.

35. Ortiz, January 10, 2007.

36. Ibid.

37. Green, "Race and the Space Race."

38. Ibid.

39. Ibid.

40. Ibid.

41. Ortiz, January 10, 2007.

42. Montgomery, February 6, 2008.

43. Ibid.

44. Julius Montgomery, conversation with Richard Paul, February 24, 2011.

45. John F. Kennedy, "Moon Speech" (speech), Rice Stadium, Rice University, Houston, TX, September 12, 1962, http://er.jsc.nasa.gov/seh/ricetalk.htm.

46. "President Eisenhower Delivers Farewell Address to the Nation," *NBC News*, New York, NBC Universal, January 17, 1961, https://archives.nbclearn.com/portal/site/k-12/browse/?cuecard=60883.

47. "President Eisenhower Delivers Farewell Address to the Nation." https://archives.nbclearn.com/portal/site/k-12/browse/?cuecard=60883.

48. Howard McCurdy, conversation with Richard Paul, April 9, 2009.

49. Peter Kuznick, conversation with Richard Paul, July 26, 2005. Kuznick is director of the Nuclear Studies Institute at American University, Washington, DC.

50. Paul Boyer, conversation with Richard Paul, August 4, 2005. Boyer is the author of *By the Bomb's Early Light: American Thought and Culture at the Dawn of the Atomic Age* (Chapel Hill: University of North Carolina Press, 1994).

51. Alan Winkler, conversation with Richard Paul, July 30, 2005. Winkler is a professor of history at Miami University–Ohio and author of *Life under a Cloud: American Anxiety about the Atom* (Champaign: University of Illinois Press, 1993).

52. Roger Launius, conversation with Richard Paul, March 13, 1998. Launius is a curator at the National Air and Space Museum.

53. Margaret Weitekamp, conversation with Richard Paul, August 1, 2007. Weitekamp is a curator at the National Air and Space Museum.

54. Howard McCurdy, conversation with Richard Paul, April 3, 2009.

55. George Reedy, "The Legislative Origins of the Space Act: Proceedings of a Video-taped Workshop," NASA Space Act Origins—1992 Symposium hosted by GWU/LBJL, April 3, 1992, folder 12216, NASA Headquarters Historical Reference Collection, Washington, DC.

56. "President Eisenhower Touts U.S. Scientific Advancements," Universal news-reel, New York, NBC Universal, November 7, 1957, https://archives.nbclearn.com/portal/site/k-12/browse/?cuecard=1618.

57. Meena Bose, conversation with Richard Paul, April 7, 2009. Bose is Peter S. Kalikow Chair in Presidential Studies at Hofstra University.

58. Howard E. McCurdy, conversation with Richard Paul, July 8, 1999.

59. "Lyndon Johnson Speech to the Anti-Defamation League of B'nai B'rith," *CBS News*, New York, CBS Television Network, October 20, 1957, Film Serial CBS-C1R2, LBJL.

60. Ibid.

61. The President's News Conference, October 9, 1957, http://www.presidency.ucsb.edu/ws/?pid=10924.

62. "Lyndon Johnson Speech to the Anti-Defamation League."

63. The President's News Conference, February 17, 1960, http://www.presidency.ucsb.edu/ws/?pid=12039.

64. McCurdy, April 3, 2009.

65. Reedy, "The Legislative Origins of the Space Act."

66. U.S. Senate Preparedness Investigating Subcommittee, Committee on Armed Services, Inquiry into Satellite and Missile Programs, Part 1, Hearing, November 25, 1957, 85th Cong., 1st Sess. (Washington, DC: Government Printing Office), 3.

67. Walter A. McDougall, . . . the Heavens and the Earth: A Political History of the Space Age (New York: Basic Books, 1985), 151–153.

68. Ibid., 151.

69. Allen J. Matusow, *The Unraveling of America: A History of Liberalism in the 1960s* (New York: Harper and Row, 1984), 10–11.

70. Roger Launius, conversation with Richard Paul, April 10, 2009.

71. Ibid.

72. Roger E. Bilstein, *Orders of Magnitude: A History of the NACA and NASA, 1915–1990*, 3rd ed. (Washington, DC: Government Printing Office, 1989). Bilstein's book is one of several NASA histories placed online by the NASA History Office. http://www.hq.nasa.gov/office/pao/History/SP-4406/chap.3.html.

73. Reedy, "Legislative Origins of the Space Act."

74. "President Eisenhower Delivers Farewell Address to the Nation." https://archives.nbclearn.com/portal/site/k-12/browse/?cuecard=60883.

75. Bruce Schulman's book *Cotton Belt to Sunbelt* is the most thorough work outlining the impact of federal spending on the South.

76. Gordon Patterson, "Countdown to College: Launching Florida Institute of Technology," *Florida Historical Quarterly* 77, no. 2 (Fall 1998): 175.

77. Ibid., 163.

78. Ibid., 170.

79. Ibid., 175.

80. Montgomery, February 6, 2008.

81. Raymond A. Bauer, Richard S. Rosenbloom, and Laure Sharp, *Second-Order Conse-quences: A Methodological Essay on the Impact of Technology* (Cambridge, MA: MIT Press, 1969), 96. Bauer et al. quote the superintendent of schools from Peter Dodd's "Social Change in Space-Impacted Communities," a document of the Committee on Space of the American Academy of Arts and Sciences (Cambridge, MA, August 1964), 36; William S. Ellis, "Space Crescent II: The Brain Ghettos," *Nation*, October 19, 1964, 243.

82. Patterson, "Countdown to College," 175.

83. Montgomery, February 6, 2008.

84. Patterson, "Countdown to College," 175.

85. Montgomery, February 6, 2008.

86. *U.S. News and World Report 2012–2013 College Rankings*, http://colleges.usnews .rankingsandreviews.com/best-colleges/florida-tech-1469.

87. Patterson, "Countdown to College," 175.

88. Button, *Blacks and Social Change*, 9, 71.

89. Susanne Cervenka, "Melbourne City Council Vote May End Diversity Drought: City Hasn't Seen Black Council Member since 1977 Elections," *Florida Today*, November 3, 2012, http://www.floridatoday.com/article/20121103/NEWS05/311030019/Melbourne-city -council-vote-may-end-diversity-drought.

Chapter 2

1. Ray, February 16, 2009.

2. Samuel P. Huntington, *Who Are We? The Challenges to America's National Identity* (New York: Simon and Schuster), 2005, 54; Bernie D. Jones, "Southern Free Women of Color in the Antebellum North: Race, Class, and a 'New Women's Legal History,'" *Akron Law Review* 41, no. 3 (2007–2008): 764; Ayana Byrd and Lori Tharps, *Hair Story: Untangling the Roots of Black Hair in America* (New York: St. Martin's, 2002), 22; Elizabeth Stone, *Black Sheep and Kissing Cousins: How Our Family Stories Shape Us* (Edison, NJ: Transaction, 2004), 115.

3. Marshall Space Flight Center Manpower Office, "A Chronology of the Equal Employ-ment Opportunity Program at MSFC," February 1971, p. 1, Equal Opportunity Employment series, NASA Marshall Space Flight Center Historical Reference Collection, Huntsville, AL.

4. Ibid.

5. Peter C. Dodd, "The Slow Pace of Change in the Space-Centered Communities: A Report on Technology and Modern Society," *Cahiers d'historie Mondiale, Journal of World History* 10, no. 3 (1967): 570.

6. Ray, February 16, 2009.

7. Roz Foster, conversation with Richard Paul, January 30, 2008.

8. Susan Parker and Robert W. Blythe, eds., *Canaveral National Seashore Historic Re-source Study* (Washington, DC: Government Printing Office, 2008), 77, http://www.nps .gov/history/history/online_books/cana/cana_hrs.pdf.

9. Ibid.

10. According to a sign placed at the site of the schoolhouse by the Brevard County His-torical Commission and the Brevard County Tourist Development Council, "Campbell's chil-dren included Florida, Eugenia, Agnes, Henry and Willie, who was Valedictorian in 1892. Jackson's children were Annie, Mary, Floyd and Douglas, who was Valedictorian in 1893."

11. Parker and Blythe, *Canaveral*, 78.

12. Foster, January 30, 2008.

13. Carla Kaplan, ed., *Zora Neal Hurston: A Life in Letters* (New York: Anchor Books, 2002), 756.

14. Ray, February 16, 2009.

15. Letter, James E. Webb to Robert F. Kennedy, June 6, 1963, James E. Webb Files, NASA Historical Reference Collection, NASA Headquarters, Washington, DC.

16. Ray, February 16, 2009.

17. Parker and Blythe, *Canaveral*, 78.

18. Foster, January 30, 2008.

19. Ray, February 16, 2009.

20. Luther Hodges, "What Kind of America," in *The Deep South in Transformation: A Symposium*, ed. Robert B. Highsaw (Tuscaloosa: University of Alabama Press, 1964), 30, 42.

21. On July 1, 1962, the Cape Canaveral Missile Test Annex was renamed the Florida Launch Operations Center (LOC). Marshall Space Flight Center (MSFC) in Alabama originally administered the LOC. The Kennedy Space Center archives have no records regarding the center's equal employment opportunity (EEO) activities in the early 1960s. Researchers can find what LOC EEO material there is in the MSFC archive. There are also a handful of documents in the NASA History Office in a folder devoted to NACA/NASA Equal Employment Opportunity Programs, 1923–1992.

22. Ben Hursey, personnel officer, *John F. Kennedy Space Center*, NACA/NASA Equal Employment Opportunity Programs 1923–1992, file 188977, NASA EEO, e. 1960s Federal Records Center, NASA Headquarters Historical Reference Collection, Washington, DC.

23. Theodis Ray, conversation with Richard Paul, May 2, 2013.

24. Report on Evaluation of Personnel Management Activities, LOC, September 16–27, 1963, NACA/NASA Equal Employment Opportunity Programs 1923–1992, file 188977, NASA EEO, e. 1960s Federal Records Center, NASA Headquarters Historical Reference Collection, Washington, DC.

25. During the search that brought us to Julius Montgomery, Richard Paul contacted numerous NASA retiree groups in Florida. No one interviewed could recall ever seeing African American workers at Cape Canaveral. It seems a reasonable assumption that if whites did not even know that blacks were there, they were not aware that blacks had grievances.

26. "Results of EEO Counseling Questionnaires, Agency-Wide," 1963, NACA/NASA Equal Employment Opportunity Programs 1923–1992, file 188977, NASA Headquarters Historical Reference Collection, Washington, DC.

27. *A Chronology of the Equal Employment Opportunity Program at MSFC*, Manpower Office, Administration and Technical Services, George C. Marshall Space Flight Center, 1971, Equal Opportunity Employment series, NASA Marshall Space Flight Center Historical Reference Collection, Huntsville, AL, 5.

28. Letter, James E. Webb to Floyd L. Thompson, December 12, 1961, James E. Webb Files, NASA Historical Reference Collection, NASA Headquarters, Washington, DC. A March 12, 1962, memo written at MSFC suggests the compliance officer at the other centers received the same letter.

29. Letter, James E. Webb to Wernher von Braun, June 24, 1963, Minority Groups (1961–1993), file 008983, NASA Headquarters Historical Reference Collection, Washington, DC.

30. *NASA Contracts Equal Employment Opportunity Program Achievement during 1963*, NACA/NASA Equal Employment Opportunity Programs 1923–1992, file 188977, NASA EEO, e. 1960s Federal Records Center, NASA Headquarters Historical Reference Collection, Washington, DC.

31. Montgomery, February 6, 2008.

32. Ray, February 16, 2009.

33. Ibid.

34. Bill Bell, conversation with Richard Paul, February 4, 2008. Bell has been mayor of Durham, North Carolina, since 2001.

35. John F. Kennedy, "Special Message to the Congress on Urgent National Needs, May 25, 1961," John F. Kennedy Presidential Library and Museum (hereafter JFKL), Boston, MA, http://www.jfklibrary.org/Research/Ready-Reference/JFK-Speeches/Special-Message-to -the-Congress-on-Urgent-National-Needs-May-25-1961.aspx.

36. C. Vann Woodward, *The Strange Career of Jim Crow* (New York: Oxford University Press, 1974), 169.

37. Packard, *American Nightmare*, 262–263.

38. Woodward, *Strange Career*, 171.

39. James Hilty, *Robert Kennedy: Brother Protector* (Philadelphia: Temple University Press, 1997), 320.

40. Anecdotes appear in many civil rights histories that have the president speaking this line to speechwriter Harris Wofford, though it is not clear about whom he was complaining. Robert Dallek in *An Unfinished Life: John F. Kennedy, 1917–1963* and Richard Reeves in *President Kennedy: Profile of Power*, among others, say he made the comment after he had been criticized at a public function by Harry Belafonte for not doing more to help the Freedom Riders in 1961. Robert Dallek, *An Unfinished Life: John F. Kennedy, 1917–1963* (New York: Little, Brown, 2003); Richard Reeves, *President Kennedy: Profile of Power* (New York: Simon and Schuster, 1994). Nick Bryant says the president was responding to criticism from Yale Law School dean Eugene Rostow, who was at the function with Belafonte. Nick Bryant, *The Bystander: John F. Kennedy and the Struggle for Black Equality* (New York: Basic Books, 2006).

41. Samuel Leiter and William Leiter, "Affirmative Action and the Presidential Role in Modern Civil Rights Reform: A Sampler of Books on the 1990s," *Presidential Studies Quarterly* 29, no. 1 (March 1999): 180–181.

42. Graham, *Civil Rights Era*, 79.

43. Telephone conversation, President John F. Kennedy and Governor Ross Barnett, September 29, 1962, 2:00 p.m., http://soundlearning.publicradio.org/subjects/history_civics /whitehouse_on_civilrights/Transcript_%20JFK%20Talks%20with%20Mississip.pdf.

44. Hilty, *Robert Kennedy*, 315.

45. Golland, *Constructing Affirmative Action*, 37.

46. MacLean, "From the Benighted South."

47. Michael R. Beschloss, "Kennedy and the Decision to Go to the Moon," in *Spaceflight and the Myth of Presidential Leadership* (Chicago: University of Illinois Press, 1997), 51.

48. Howard E. McCurdy, conversation with Richard Paul, July 8, 1999.

49. Mary L. Dudziak, *Cold War Civil Rights: Race and the Image of American Democracy* (Princeton, NJ: Princeton University Press, 2011), 153.

50. Beschloss, *Kennedy*, 52.

51. McCurdy, July 8, 1999,

52. Beschloss, *Kennedy*, 56.

53. Howard McCurdy, quoted in "Washington Goes to the Moon," Public Radio Exchange, http://www.prx.org/pieces/629/transcripts/629.

54. U.S. Department of Commerce, *1960 Census of the Population, Vol. 1: Characteristics of the Population, Pt. 11: Florida* (Washington, DC: Government Printing Office, 1963), 11–12, 11–90; *1970 Census of the Population, Vol. 1: Characteristics of the Population, Pt. 11: Florida* (Washington, DC: Government Printing Office, 1973), 11–149.

55. Ellis, "Space Crescent II," 241–242.

56. Ibid.

57. Ibid., 242.

58. Ibid., 241–242.

59. Ibid., 241.

60. Letter, Roy Wilkins to David Lawrence, March 15, 1963, NK-15, microfilm records of the Housing and Home Finance Agency, roll 8, JFKL; letter, Roy Wilkins to John F. Kennedy, March 18, 1963, "HU 2/ST 2-HU 2/ ST 9" folder ("HU 2 General"), White House Central Subject File (WHCSF), box 368, JFKL; Ellis, "Space Crescent II," 241; "NAACP Charges Bias at Cape Canaveral," *Atlanta Daily World*, March 24, 1963, A1, col. 1; "Canaveral Color Bars Are Bared," *Baltimore Afro-American*, March 30, 1963, 1, 2.

61. Letter, Wilkins to Kennedy, March 18, 1963.

62. Letter, Alfred S. Hodgson to Lee C. White, April 5, 1963, "HU 2/ST 2-HU 2/ ST 9" folder ("HU 2 General"), WHCSF, box 368, JFKL; letter, Lee C. White to Roy Wilkins, March 22, 1963, "HU 2/ST 2-HU 2/ ST 9" folder ("HU 2 General"), WHCSF, box 368, JFKL; letter, Alfred S. Hodgson to Lee C. White, April 5, 1963, "HU2/ST1-HU2/ST9" folder ("HU2 General"), WHCSF, box 368, JFKL; "Cape Canaveral Bias under Probe," *Pittsburgh Courier*, May 11, 1963, 6, col. 6.

63. Ray, May 2, 2013.

64. Ray, February 16, 2009.

65. Ray, May 2, 2013.

66. Bauer et al., *Second-Order Consequences*, 67.

67. Ibid., 75, 98.

Chapter 3

1. "Missile Scientist on Target," *Ebony*, September 1975, 158.

2. Frank Crossley, conversation with Richard Paul, November 12, 2008.

3. Roger D. Launius, "American Memory, Culture Wars, and the Challenge of Presenting Science and Technology in a National Museum," *Public Historian* 29, no. 1 (Winter 2007): 14.

4. Hayden White, *Metahistory: The Historical Imagination in Nineteenth-Century Europe* (Baltimore: Johns Hopkins University Press, 1973), ix.

5. Jennifer Delton, "Before the EEOC: How Management Integrated the Workplace," *Business History Review* 81, no. 2 (Summer 2007): 269.

6. "Missile Scientist on Target," 158.

7. Crossley, November 12, 2008.

8. S. I. Hayakawa, "Second Thoughts: Solitary Negro Student," *Chicago Defender* (national edition), February 3, 1945, 11.

9. Joseph P. Reidy, "Black Men in Navy Blue During the Civil War," *Prologue* 33, no. 3 (Fall 2001), http://www.archives.gov/publications/prologue/2001/fall/black-sailors-1.html.

10. Joseph Reidy, conversation with Richard Paul, October 24, 2001.

11. Mark Clague, conversation with Richard Paul, October 18, 2011. Dr. Clague is Associate Professor of Music, American Culture, and African American Studies, University of Michigan.

12. Alex Albright, conversation with Richard Paul, October 17, 2011. Albright is the writer and producer of *Boogie in Black and White* (c. 1988), Boogie in Black and White Documentary Collection (#1086), East Carolina Manuscript Collection, J. Y. Joyner Library, East Carolina University, Greenville, NC.

13. Hayakawa, "Second Thoughts," 11.

14. Ibid.

15. Crossley, November 12, 2008.

16. Hayakawa, "Second Thoughts," 11.

17. Ibid.

18. Robert Dallek, "Johnson, Project Apollo, and the Politics of Space Program Planning," in *Spaceflight and the Myth of Presidential Leadership* (Chicago: University of Illinois Press, 1997), 72.

19. Schulman, December 2, 2008.

20. Gavin Wright, "The Economic Revolution in the American South," *Economic Perspectives* 1, no. 1 (Summer 1987): 170.

21. David E. Bernstein, *Only One Place of Redress: African Americans, Labor Regulations, and the Courts from Reconstruction to the New Deal* (Durham, NC: Duke University Press), 2001, 10; MacLean, "From the Benighted South," 215; transcript, Robert S. McNamara Oral History, Special Interview I, March 26, 1993, by Robert Dallek, electronic copy, LBJL, 4.

22. Wilbur Joseph Cash, *The Mind of the South* (New York: Vintage Books, 1960).

23. C. Vann Woodward and Walter A. McDougall have both referred to this period as the "Second Reconstruction." McDougall, . . . *the Heavens*, 376; Woodward, *Strange Career*, 8.

24. Allen J. Matusow, *The Unraveling of America: A History of Liberalism in the 1960s* (New York: Harper and Row, 1984), 61.

25. Kevin J. McMahon, *Reconsidering Roosevelt on Race: How the Presidency Paved the Road to Brown* (Chicago: University Of Chicago Press, 2003), 100.

26. Bernstein, *Only One Place of Redress*, 10.

27. MacLean, "From the Benighted South," 213; Matusow, *Unraveling of America*, 60–61; Graham, *Civil Rights Era*, 9.

28. Woodward, *Strange Career*, 134.

29. Dallek, "Johnson, Project Apollo," 68.

30. McCurdy, July 8, 1999.

31. Jennings, November 20, 2008.

32. Crossley, November 12, 2008.

33. Employees who held the job titled "computer" made (by tabulating machine, by hand, and by slide rule) the millions of mathematical calculations that the processing chips in the common desktop computer make today.

34. Woodward, *Strange Career*, 9.

35. Graham, *Civil Rights Era*, 29.

36. Woodward, *Strange Career*, 132–133.

37. Jerrold M. Packard, *American Nightmare*, 228–229.

38. Woodward, *Strange Career*, 133.

39. Matusow, *Unraveling of America*, 63.

40. Graham, *Civil Rights Era*, 38.

41. Matusow, *Unraveling of America*, 64.

42. Address of Secretary Wirtz at the Michigan celebration of the hundredth anniversary of the Emancipation Proclamation, Detroit, MI, June 30, 1963.

43. Martin Luther King Jr., "Equality Now," *Nation*, February 4, 1961, 93.

44. Graham, *Civil Rights Era*, 24.

45. N. Thompson Powers, "Federal Procurement and Equal Employment Opportunity," *Law and Contemporary Problems* 29, no. 2 (Spring 1964): 473.

46. "For Negroes: More and Better Jobs in Government," *U.S. News and World Report*, March 5, 1962, 83–85.

47. Executive Order number 10925, 26 *Fed. reg.* 1977, March 8, 1961.

48. Judson MacLaury, "President Kennedy's E.O. 10925: Seedbed of Affirmative Action," *Federal History (Society for History in the Federal Government)* 2 (January 2010): 57.

49. Crossley, November 12, 2008.

50. Delton, "Before the EEOC," 272–273.

51. Jennifer Delton, *Racial Integration in Corporate America, 1940–1990* (London: Cambridge University Press, 2009), 117.

52. Delton, "Before the EEOC," 277.

53. Ibid., 271.

54. "NOW . . . You Can Be a Part!" (Thiokol Chemical Corporation, advertisement), *Houston Post*, October 16, 1961, 14; "Success of First Saturn Flight . . ." (General Electric, advertisement), *Houston Post*, December 3, 1961, 13.

55. Graham, *Civil Rights Era*, 40.

56. Ibid., 27.

57. MacLaury, "President Kennedy's E.O. 10925," 46.

58. Bryant, *Bystander*, 230.

59. Letter, James E. Webb to Lyndon Johnson, April 13, 1961, NACA/NASA Equal Employment Opportunity Programs 1923–1992, file 188977, NASA EEO, e. 1960s Federal Records Center, NASA Headquarters Historical Reference Collection, Washington, DC.

60. Letter, James E. Webb to Floyd L. Thompson, December 12, 1961, James E. Webb Files, NASA Historical Reference Collection, NASA Headquarters, Washington, DC.

61. James E. Webb, Memo to All Program Directors and Staff Officers, Headquarters; All Directors of Field Installations, April 11, 1961, NACA/NASA Equal Employment Opportunity Programs 1923–1992, file 188977, NASA EEO, e. 1960s Federal Records Center, NASA Headquarters Historical Reference Collection, Washington, DC.

62. "New Compliance Officer," *Marshall Star*, May 10, 1961, 8, Equal Opportunity Employment series, Digital Media, NASA Marshall Space Flight Center Historical Reference Collection, Huntsville, AL.

63. Letter, Wernher von Braun to James Webb, undated but makes reference to "your letter of December 12, 1961," Equal Opportunity Employment series, Digital Media, NASA Marshall Space Flight Center Historical Reference Collection, Huntsville, AL.

64. Graham, *Civil Rights Era*, 47.

65. "Half a Million Workers," *Fortune*, no. 3, March 1941, 98, 163.

66. "Newest Aircraft Factory Built by Government Funds—Closed Tight to Negroes," *Philadelphia Tribune*, March 27, 1941, 2, col. 3.

67. "Boeing Aircraft Says No Place for Negroes," *Norfolk New Journal and Guide*, December 13, 1941, 4, col. 1.

68. Kevin Allen, *The Battle for Los Angeles: Racial Ideology and World War II* (Albuquerque: University of New Mexico Press, 2006), 44.

69. Herbert R. Northrup, *The Negro in the Aerospace Industry* (Philadelphia: University of Pennsylvania Press, 1968), 24, 31, 33.

70. Ibid., 22; Graham, *Civil Rights Era*, 47–48.

71. "Urges U.S. Kill Order with Biased Ga. Plant," *Chicago Defender*, April 4, 1961, 4.

72. According to Graham, NAACP labor secretary Herbert Hill "extracted from [assistant secretary of labor] Jerry Holloman a public pledge to cancel the contract of any employer who refused to comply with president's new ban on discrimination." Graham, *Civil Rights Era*, 48.

73. Northrop, *Aerospace Industry*, 76.

74. Graham, *Civil Rights Era*, 51–54.

75. Powers, "Federal Procurement," 176.

76. Golland, *Constructing Affirmative Action*, 45, 48.

77. President's Committee on Equal Employment Opportunity, July 19, 1963, 26, LBJL.

78. Powers, "Federal Procurement," 480.

79. Golland, *Constructing Affirmative Action*, 42.

80. Douglas Helms, "Eroding the Color Line: The Soil Conservation Service and the Civil Rights Act of 1964," *Agricultural History* 65, no. 2 (Spring 1991): 47.

81. Hilty, *Robert Kennedy*, 300–302; Graham, *Civil Rights Era*, 70–71.

82. Houston Chapter of the Episcopal Society for Cultural and Racial Unity, *Nowhere to Go: A Study of the Plight of Negroes in Houston, Texas*, 8, HU2/ST42-ST50, HU, box 41, LBJL.

83. Northrup, *Aerospace Industry*, 31.

84. Hayakawa, "Second Thoughts," 11.

85. "Will Get First Engineering Ph.D. at Ill. Tech," *Chicago Defender*, June 17, 1950, 23.

Chapter 4

1. Andrew D. Grossman, "The Early Cold War and American Political Development: Reflections on Recent Research," *International Journal of Politics, Culture, and Society* 15, no. 3 (Spring 2002): 473.

2. A number of books focus on the state of mind of important American intellectual elites during the early years of the Cold War, including Michael Hogan, *A Cross of Iron: Harry S. Truman and the Origins of the National Security State, 1945–1954* (Cambridge University Press, 1998), and Guy Oakes, *The Imaginary War: Civil Defense and American Cold War Culture* (Oxford University Press, 1994).

3. Richard Hofstadter, *The Progressive Historians: Turner, Beard, Parrington* (New York: Alfred A. Knopf, 1968), 438.

4. Peter Charles Hoffer, *Past Imperfect: Facts, Fictions, Fraud—American History from*

Bancroft and Parkman to Ambrose, Bellesiles, Ellis, and Goodwin (New York: Public Affairs, 2004), 63.

5. Karsten Werth, "A Surrogate for War—The U.S. Space Program in the 1960s," *Amerikastudien/American Studies* 49, no. 4 (2004): 563.

6. *Hold On! (There's No Place Like Space)*, directed by Arthur Lubin (1966; Hollywood: Metro-Goldwyn-Mayer); *Way . . . Way Out*, directed by Gordon Douglas (1966; Hollywood: Twentieth Century-Fox) (for film poster, see http://www.britposters.com/images/way%20 way%20out%20320x240.jpg); *The Reluctant Astronaut*, directed by Edward J. Montagne (1967; Hollywood: Universal Pictures).

7. *The Jetsons*, http://www.classictvhits.com/show.php?id=318; *Lost in Space*, http:// www.classictvhits.com/show.php?id=534; *Star Trek*, http://www.imdb.com/title/tt0060028/; *I Dream of Jeannie*, http://en.wikipedia.org/wiki/I_Dream_of_Jeannie.

8. Frederick J. Baskaw, *The Dynamic American City* (ca. 1956), http://archive.org/details /DynamicA1956_2.

9. American Institute of Architects, *No Time for Ugliness* (Washington, DC: AIA, 1965), http://archive.org/details/no_time_for_ugliness_1.

10. Wayne Thompson in *Conference on Space, Science, and Urban Life: Proceedings of a Conference Held at Oakland, California, March 28–30, 1963*, National Aeronautics and Space Administration, vol. 1 (Washington, DC: Office of Scientific and Technological Information, National Aeronautics and Space Administration, 1963).

11. Ibid., 2.

12. "Patterson, Faubus Criticize Kennedy and Justice Dept.," *News and Courier*, October 2, 1962, 8-A.

13. "Governors Keep Eye on Mississippi," *Danville Bee*, October 1, 1962, 7.

14. Lyndon B. Johnson, "The New World of Space" (speech), *Proceedings of the Second National Conference on the Peaceful Uses of Space*, Seattle, WA, May 8–10, 1962 (Washington, DC: Government Printing Office, 1962), 30.

15. UPI, "Astronaut Grissom Rarin' to Go," *Daily Defender*, July 18, 1961, 4, col. 1.

16. Jack Hicks, "Race and Space," *New York Amsterdam News*, April 16, 1961, 12, col. 7.

17. George M. Coleman, "Spaceman Survives Tough One:Hinesville Negroes Lose Out," *Atlanta Daily World*, May 25, 1962, 1, col. 5.

18. Ibid.

19. Technically, the first words spoken by Neil Armstrong on the Moon were "contact light," but his first sentence spoken back to Earth was "Houston, Tranquility Base here; the Eagle has landed."

20. Houston Chapter, *Nowhere to Go*, 8.

21. "10 Best Cities for Negro Employment," *Ebony*, March 1965, 120.

22. Otis King, conversation with Richard Paul, December 24, 2008.

23. Charles Zelden, conversation with Richard Paul, January 9, 2007. Zelden is author of *Battle for the Black Ballot: Smith V. Allwright and the Defeat of the Texas All-White Primary* (Lawrence: University Press of Kansas, 2004).

24. King, December 24, 2008.

25. Fritz Lanham, "The Silent End of Segregation: The Behind-the-Scenes Story of Houston's Peaceful Integration in the 60s," *Houston Chronicle*, June 15, 1997, 8.

26. Quentin Mease, conversation with Richard Paul, January 7, 2009.

27. Zelden, January 9, 2007; King, December 24, 2008; Quentin Mease, *On Equal Footing* (Austin, TX: Eakin Press), 2001, 93, 95.

28. "Texas Firm Wants Contract but Rejects Equality Clause," *Baltimore Afro-American*, July 22, 1961, 1; Taylor Branch, *Parting the Waters: America in the King Years, 1954–63* (New York: Simon and Schuster, 1988), 86. Branch quoted Johnson's account of the incident from a July 9, 1963, White House meeting.

29. David Allerd, "Chances Are 99–1 That Houston Will Get the Proposed Space Lab," *Houston Post*, August 24, 1961, 1(1); Ralph O'Leary, "NASA Wants Test Site Near Space Laboratory," *Houston Post*, August 26, 1961, p. 1, sec. 1.

30. Allerd, "Chances Are 99–1," p. 1, sec. 1.

31. Ralph S. O'Leary, "NASA to Build Space Center on Clear Lake," *Houston Post*, September 20, 1961, 1.

32. David G. McComb, *Houston: A History* (Austin: University of Texas Press), 1981, 142; O'Leary, "NASA to Build," 1.

33. Fredericka Meiners, *A History of Rice University* (Houston, TX: Rice University Studies Special Publications, 1982), http://archive.org/stream/historyofriceuni00mein /historyofriceuni00mein_djvu.txt.

34. Melissa Fitzsimons Kean, "At a Most Uncomfortable Speed: The Desegregation of the South's Private Universities, 1945–1964" (PhD diss., Rice University, 2000), 413, 417.

35. Ibid., 412.

36. Ibid., 417.

37. Ibid., 419, 422, 423.

38. Ibid., 426.

39. Ibid., 434.

40. Marshall Verniaud, "Pitzer Says Rice Future Tied to End of Race Bars," *Houston Post*, February 18 1964, 1, 7, col. 1; Marshall Verniaud, "Restrictions Hinder Rice, Jury Decides," *Houston Post*, February 22, 1964, 1, col. 1; Marshall Verniaud, "Ruling Will Permit Rice Tuition, End Race Bars," *Houston Post*, March 10, 1964, 1, 6, col. 1; McComb, *Houston*, 174.

41. King, December 24, 2008.

42. Northrup, *Aerospace Industry*, 221.

43. "Clear Lake City, TX," http://www.tshaonline.org/handbook/online/articles/hjc23; Richard D. Lyons, "Booming Houston's Eyes Are on Astronauts," *New York Times*, October 22, 1968, 35.

44. "And Now . . . Over to Houston," *Newsweek*, June 14, 1965, 36; "The Meaning of the March," *Informer (Houston)*, May 15, 1965, 4.

45. Lyons, "Booming Houston's Eyes," 35.

46. Hofstadter, *Progressive Historians*, 438.

47. George Simpson in *Conference on Space, Science, and Urban Life: Proceedings of a Conference Held at Oakland, California, March 28–30, 1963*, National Aeronautics and Space Administration, vol. 6 (Washington, DC: Office of Scientific and Technological Information, National Aeronautics and Space Administration, 1963).

48. Meeting tape 63A, JFKL, November 21, 1962, from JFK and the Space Race, Presidential Recordings Program, Miller Center of Public Affairs, University of Virginia, http:// millercenter.org/presidentialclassroom/exhibits/jfk-and-the-space-race.

49. Enoc P. Waters Jr., "They Helped Track Glenn in Orbit," *Baltimore Afro-American*, March 3, 1962, 1; UPI, "Cooper Relays 'Hello' to Africa from Space Craft," *Chicago Defender*,

May 18, 1963, 3, col. 1; Nicholas J. Cull, *The Cold War and the United States Information Agency* (London: Cambridge University Press, 2009), 212.

50. Teasel Muir-Harmony, "From Spacecraft to Icon: Friendship 7's 'Fourth Orbit,'" unpublished manuscript (2012).

51. Walter Cronkite, conversation with Richard Paul, February 26, 1999.

52. Richard Witkin, "Cooper Is Flying Smoothly On, Will Try for at Least Seventeen Orbits: He Ejects Beacon, Then Sleeps," *New York Times*, May 16, 1963, 1; Richard Witkin, "Cooper Maneuvers to a Bullseye Landing with Manual Control as as Automatic Fails: 'I'm in Fine Shape,' He Says after Twenty-Two Orbits," *New York Times*, May 17, 1963, 1; John W. Finney, "Cooper Hailed in Capital: He Will Come Here Today," *New York Times*, May 22, 1963, 1; Foster Hailey, "City Roars Big 'Well Done' to Cooper: Throngs Greet Astronaut at Parade and Luncheon," *New York Times*, May 23, 1963, 1.

53. Mease, January 7, 2009.

54. Gregory Curtis, "The First Protestor," *Texas Monthly*, 1997, http://www.gregorycurtis.com/greg-art5.htm.

55. Mease, *On Equal Footing*, 94.

56. King, December 24, 2008; Mease, *On Equal Footing*, 93, 94.

57. Lanham, "The Silent End of Segregation," 8.

58. King, December 24, 2008.

59. Mease, January 7, 2009.

60. Mease, *On Equal Footing*, 98.

61. Lanham, "The Silent End of Segregation," 8.

62. Ibid.

63. Debra Ann Reid, *Inalienable Rights: Texans and Their Quests for Justice* (College Station: Texas A&M University Press 2009), 128.

64. *1964 Staff Report: Public Education*, 223–224; Harry K. Wright, *Civil Rights U.S.A.: Public Schools: Southern States, 1963: Texas* (Washington, DC: Government Printing Office, 1964), 34, 72; "The Meaning of the March," 4.

65. King, December 24, 2008.

66. Letter, Lee C. White to Mildred L. Pierce, Elizabeth Chaupette, John Gilbride, and Robert Reed, October 12, 1962; Route slip, White House to NASA regarding October 12, 1962; letter, A. N. Feldzamen to Kenneth O'Donnell, October 16, 1962, "HU 9-1-62.11–30–62" folder ("HU2 General"), WHCSF, box 362, JFKL.

67. "Congratulations to Astronaut John Glenn," *Atlanta Daily World*, February 22, 1962, 4, col. 1.

68. "The Negro and Space Agency," *Daily Defender*, March 1, 1962, 11, col. 1.

69. Dan Burley, "A Negro Astronaut? You Can Forget That Thought!" *Philadelphia Tribune*, August 12, 1961, 4, col. 6.

Chapter 5

1. "Select Negro for Aerospace School: May Become First of Race in Space," *Cleveland Call and Post*, July 7, 1962, 1C, col. 3.

2. "Report on First Negro Astronaut Trainee," *Jet*, April 18, 1963, cover; "America Trains First Negro Spaceman," *Sepia*, June 1963, cover; "KCK Native First Negro Astronaut," *Kansas City Call*, no date, from personal collection of Edward Dwight; "New Astronaut May

Be First Man on Moon: Negro Astronaut Trainee May Be First American to Reach Moon," *Daily Defender*, April 1, 1963, 1, col. 1.

3. Joseph D. Atkinson Jr. and Jay M. Shafritz, *The Real Stuff: A History of NASA's Astronaut Recruitment Program* (New York: Praeger, 1985), 98–100.

4. U.S. Air Force training programs at Edwards Air Force Base underwent several name changes between 1952 and 1961. This can be confusing. From 1952 to 1955 the name was the U.S. Air Force Experimental Flight Test Pilot School. In 1955 it was renamed the U.S. Air Force Flight Test Pilot School. The program expanded to include astronaut training in October 1961, becoming the U.S. Air Force Aerospace Research Pilot School (ARPS). ARPS was a full-year program with two elements: Phase I: Experimental Test Pilot Course, and Phase II: Aerospace Research Pilot Course (ARPC). The school changed names again in 1972. The designations used by White House staff, U.S. Air Force personnel, and media reflect these changes between 1961 and 1965. http://www.edwards.af.mil/library/factsheets/factsheet.asp?id=6586.

5. Memoranda, Frederick Dutton to Adam Yarmolinsky, August 1, 1961; Adam Yarmolinsky to Frederick Dutton, August 18, 1961, "HU 9-8-61→9-30-61" folder ("HU2 General"), WHCSF, box 362, JFKL.

6. Memoranda, Frederick Dutton to Adam Yarmolinsky, August 23, 1961, "HU 9-8-61→9-30-61" folder ("HU2 General"), WHCSF, box 362, JFKL; Adam Yarmolinsky to Frederick Dutton, November 4, 1961, "1-18-61→1-25-62" folder ("Outer Space"), WHCSF, box 652, JFKL.

7. The U.S. Air Force's Space Research Pilot Course required that an entrant be serving on active duty in the grade of lieutenant colonel or below, hold a currently effective aeronautical rating of pilot and be currently on flying status as a pilot, have a minimum of two thousand hours total flying time (including in jet aircraft), have a BA or equivalent in engineering, a physical science, or mathematics, be a graduate of the Experimental Test Pilot Course or Navy Test Pilot School, and be certified as medically qualified based upon successful completion of special medical and physiological testing procedures.

8. Memorandum, H. E. Van Ness, captain, U.S. Navy, assistant director for manned spaceflight operations, Office of Manned Spaceflight, to Directors of Manned Spaceflight, July 9, 1962, Minority Groups (1961–1993), file 008983, NASA Headquarters Historical Reference Collection, Washington, DC.

9. Atkinson and Shafritz, *Real Stuff*, 98–100; letter, James E. Webb to Uriah J. Fields, March 12, 1962, "March 1962" folder, box 2, James E. Webb Personal Papers, JFKL; "NASA, Cleric Told, Selects Astronauts on Ability Only," *Baltimore Afro-American*, March 31, 1962, 3.

10. Atkinson and Shafritz say the request came from the president. Their source is an interview with Ed Dwight. Chuck Yeager wrote that General LeMay told him Attorney General Kennedy wanted "a colored in space."

11. Chuck Yeager, *Yeager: An Autobiography* (New York: Bantam, 1986), 269–270.

12. Ibid., 270.

13. Atkinson and Shafritz, *Real Stuff*, 101; Charles L. Sanders, "The Troubles of 'Astronaut' Edward Dwight," *Ebony*, March 1965, 32; "Select Negro for Aerospace School."

14. Ed Dwight, *Soaring on the Wings of a Dream: The Struggles and Adventures of the "First Black Astronaut" Candidate* (Denver: Ed Dwight Studios, 2009), 88.

15. Ibid., 89.

16. Ibid., 98.

17. Atkinson and Shafritz, *Real Stuff*, 101; "Tagged as First Tan Astronaut," *Baltimore Afro-American*, June 7, 1962, 1; "Astronaut's Parents Are So Excited," *Baltimore Afro-American*, April 6, 1963, 1.

18. Yeager, *Yeager*, 270.

19. Atkinson and Shafritz, *Real Stuff*, 101; "Negro One of Fifteen in Space Course," *New York Times*, April 1, 1963, 48; Bill Becker, "Negro Astronaut Aiming for Moon," *New York Times*, April 2, 1963, 15; "Colored Space Candidate," *Baltimore Afro-American*, April 13, 1963, 4.

20. "Report on First Negro Astronaut Trainee," cover; "America Trains First Negro Spaceman," cover; "KCK Native First Negro Astronaut"; "New Astronaut May Be"; "No Racial Barriers for Man in Space—Astronaut," *Daily Defender*, April 8, 1963, 13, col. 1.

21. Ed Dwight, e-mail correspondence with Richard Paul, October 23, 2012.

22. Ibid.

23. Charles Lang, "Equal Opportunity in Space Science," n.d. Audio from the filmstrip was provided to Richard Paul by Dr. Lang's widow, Angela Lang.

24. Sanders, "Troubles," 36; Atkinson and Shafritz, *Real Stuff*, 104.

25. Fred Powledge, *Free at Last? The Civil Rights Movement and the People Who Made It* (Boston: Little, Brown, 1991), 474–475; Branch, *Parting the Waters*, 713–725.

26. T. Stockett, "'How in the World Can We Still Keep 'Em Down?'" *Baltimore Afro-American*, April 13, 1964, 4.

27. "Ousted Astronaut Raps Air Force," *Baltimore Afro-American*, June 12, 1965, 2; "Probe of Astronaut's Charges Urged," *Baltimore Afro-American*, June 19, 1965, 1; "NASA, Pentagon Deny Charge," *Informer* (Houston), June 5, 1965, 1; "Claims Space Program Bias," *Informer* (Houston), June 5, 1965, 1; "Fulfillment of Apollo 11 Moonshot Nears End," *Informer* (Houston), July 26, 1969, 1.

28. Atkinson and Shafritz, *Real Stuff*, 101; Sanders, "Troubles," 34; Gladwin Hill, "Negro Pilot Finds Bias in Air Force," *New York Times*, June 3, 1965, 21.

29. Sanders, "Troubles," 32.

30. Ibid., 31.

31. Yeager, *Yeager*, 271–272.

32. Atkinson and Shafritz, *Real Stuff*, 101; Sanders, "Troubles," 34; Hill, "Negro Pilot," 21; Referral, White House to NASA, October 4, 1965, OS July 9, 1965, OS, WHCF, box 4, LBJL.

33. Yeager, *Yeager*, 269.

34. Alfred J. Phelps, *They Had a Dream* (New York: Presidio Press, 1994), 22–23.

35. "Dwight, Edward," Contemporary Black Biography, *Encyclopedia.com*, http://www.encyclopedia.com/doc/1G2-3099700020.html.

36. Dwight, *Soaring*, 202–203.

37. Dwight, October 23, 2012. Emphasis by Dwight.

38. Nancy J. Weiss, conversation with Richard Paul, October 23, 2012. Weiss is the author of *Whitney M. Young, Jr. and the Struggle for Civil Rights* (Princeton, NJ: Princeton University Press, 1990).

Chapter 6

1. George Carruthers, conversation with Richard Paul, June 15, 2009.

2. George Carruthers, interview with David DeVorkin, August 18, 1992, Niels Bohr Library and Archives, American Institute of Physics, College Park, MD, http://www.aip.org /history/ohilist/32485.html.

3. Ibid.

4. Ibid.

5. Ibid.

6. Carruthers, June 15, 2009.

7. Carruthers, August 18, 1992.

8. Ibid.

9. Ibid.

10. Ibid.

11. Ibid.

12. Ibid.

13. Carruthers, June 15, 2009.

14. Carruthers, August 18, 1992.

15. Nancy Grace Roman, conversation with David H. DeVorkin, senior curator, History of Astronomy and the Space Sciences, National Air and Space Museum, April 14, 2010. Dr. Roman, one of the first women in the U.S. space program, is widely known as the "Mother of the Hubble Space Telescope" for her work as the liaison between astronomers and engineers creating the groundbreaking instrument. She described her role in the construction of the Hubble this way in an interview with Richard Paul: "Astronomers knew what they wanted, and the engineers were very happy to try to provide it to them. The problem was that to a large extent, they didn't speak to one another. Engineers and scientists do not speak the same language." Her work "as a go-between, between the two groups" is widely credited with ensuring the success of the Hubble.

16. Howard E. McCurdy, *Inside NASA: High Technology and Organizational Change in the U.S. Space Program* (Baltimore: Johns Hopkins University Press, 1993), 13.

17. Sylvia Doughty Fries and John A. Greene, *Project Apollo: NASA Engineers and the Age of Apollo* (Washington, DC: NASA Scientific and Technical Information Program, 1992), 125.

18. Sitton, "Segregationists Fight," E5.

19. John Fitzgerald Kennedy, Address on Civil Rights, June 11, 1963, http://millercenter .org/president/speeches/detail/3375.

20. Fries and Greene, *Project Apollo*, xii.

21. McCurdy, *Inside NASA*, 65.

22. Ibid., 28.

23. Fries and Greene, *Project Apollo*, 125.

24. McCurdy, *Inside NASA*, 71–72.

25. Fries and Greene, *Project Apollo*, xiii.

26. Letter, James E. Webb to Wernher von Braun and Kurt H. Debus, April 19, 1963, James E. Webb Files, NASA Historical Reference Collection, NASA Headquarters, Washington, DC.

27. McCurdy, *Inside NASA*, 50.

28. Carruthers, June 15, 2009.

29. Carruthers, August 18, 1992.

30. Jennings, November 20, 2008.

31. Morgan Watson, conversation with Richard Paul, January 27, 2009.

32. *NASA Contracts Equal Employment Opportunity Program Achievement during 1963*, p. 1, NACA/NASA Equal Employment Opportunity Programs 1923–1992, file 188977, NASA EEO, e. 1960s Federal Records Center, NASA Headquarters Historical Reference Collection, Washington, DC.

33. *NASA Contracts Equal Employment Opportunity Program Achievement during 1963*, p. 2, NACA/NASA Equal Employment Opportunity Programs 1923–1992, file 188977, NASA EEO, e. 1960s Federal Records Center, NASA Headquarters Historical Reference Collection, Washington, DC.

34. Montgomery, February 6, 2008.

35. Robert A. Caro, "LBJ, the Kennedy Assassination, and Me Excerpt: Lyndon Johnson's Biographer on the Way He—and the Thirty-Sixth President—Spent November 22, 1963," *New Republic*, November 16, 2013.

36. Robert A. Caro, conversation with Terry Gross, *Fresh Air*, May 13, 2013, http://www .npr.org/2014/02/17/276530368/in-passage-caro-mines-lbjs-changing-political-roles.

37. Mease, January 7, 2009; Montgomery, February 6, 2008.

38. "Title 14—Aeronautics and Space," *Federal Register* (January 9, 1965), vol. 30, no. 6, pt. 2, pp. 301–305, reel 121, microfilm.

39. McCurdy, *Inside NASA*, 23.

40. Letter, James E. Webb to Paul Bickle, June 6, 1963, Minority Groups (1961–1993), file 008983, NASA Headquarters Historical Reference Collection, Washington, DC.

41. Henry Hearns, conversation with Richard Paul, January 8, 2013.

42. Carruthers, June 15, 2009.

43. Dixie Eliopulos, e-mail correspondence with Richard Paul, January 8, 2013.

44. Hearns, January 8, 2013.

45. Webb to Bickle, June 6, 1963.

46. John Hodgson, interview with Richard Paul and Steven Moss, December 5, 2012.

47. "Office of Business Administration Appointments," NASA press release, December 2, 1960.

48. Memorandum, James Webb to All Program Directors and Staff Officers, Headquarters; All Directors of Field Installations, April 11, 1961, "Equal Employment" folder, James E. Webb Files, NASA Historical Reference Collection, NASA Headquarters, Washington, DC.

49. Hodgson, December 5, 2012.

50. Letter, James E. Webb to Wernher von Braun, June 24, 1963, James E. Webb Files, NASA Historical Reference Collection, NASA Headquarters, Washington, DC; letter, Wernher von Braun to James Webb, July 15, 1963; memorandum, R. P. Young to Navy Commander Kenneth J. Kier, June 6, Minority Groups (1961–1993), file 008983, NASA Headquarters Historical Reference Collection, Washington, DC.

51. John Hodgson, e-mail correspondence with Richard Paul and Steven Moss, December 29, 2012.

52. Andrew J. Dunar and Stephen P. Waring, *Power to Explore: A History of Marshall Space Flight Center, 1960–1990* (Washington, DC: NASA, 1999), 121, 124.

53. Ibid., 124.

54. Ellis, "Space Crescent II," 239.

55. Sally A. Downey, "Herbert R. Northrup, Eighty-Nine, Wharton School Professor," *Philadelphia Inquirer*, October 29, 2007, http://articles.philly.com/2007-10-29/news /25232852_1_wharton-school-wharton-faculty-economics.

56. Northrup, *Aerospace Industry*, 68, 77, 72.

57. Ibid., 68.

58. Carruthers, June 15, 2009.

59. Carruthers, August 18, 1992.

60. George Carruthers, interviewed by Glen Swanson, Johnson Space Center Oral History Project, March 25, 1999, http://www.jsc.nasa.gov/history/oral_histories/participants .html.

61. Carruthers, June 15, 2009.

Chapter 7

1. Wilbur Joseph Cash, *The Mind of the South* (New York: Vintage Books, 1960).

2. Wallace, "The 1963 Inaugural Address," p. 2.

3. Dodd, "Slow Pace of Change," 571.

4. Richard Hall, conversation with Richard Paul, December 15, 2008.

5. Werhner von Braun, a German, served in the SS during World War II, a fact that was well known by the mid-1960s.

6. Paul O'Neil, "The Splendid Anachronism of Huntsville," *Fortune*, June 1962, 226; William S. Ellis, "The Space Crescent: Moon Boom," *Nation*, October 12, 1964, 215; Schulman, *From Cotton Belt to Sunbelt*, 148; "The City That Space Built," *U.S. News and World Report*, November 12, 1962, 72, 73; Erik Bergaust, *Wernher von Braun* (Washington, DC: National Space Institute, 1976), 184, 191, 192, 213; see also David S. Akens, *Rocket City, USA* (Huntsville, AL: Strode, 1959).

7. Hall, December 15, 2008.

8. Letter, Harry H. Gorman to Alfred S. Hodgson, March 14, 1964, Equal Opportunity Employment series, Digital Media, NASA Marshall Space Flight Center Historical Reference Collection, Huntsville, AL.

9. Delano Hyder, conversation with Richard Paul, December 15, 2008.

10. Jane DeNeefe, *Rocket City Rock and Soul: Huntsville Musicians Remember the 1960s* (Charleston, SC: History Press, 2011), 45.

11. Hall, December 15, 2008.

12. Hyder, December 15, 2008.

13. Sonnie Wellington Hereford III and Jack D. Ellis, *Beside the Troubled Waters: A Black Doctor Remembers Life, Medicine, and Civil Rights in an Alabama Town* (Tuscaloosa: University of Alabama Press, 2011), 28.

14. MacLean, "From the Benighted South," 208.

15. Hall, December 15, 2008.

16. Dodd, "Slow Pace of Change," 572.

17. Hereford and Ellis, *Beside the Troubled Waters*, 27.

18. Ibid., 88.

19. Ibid.

20. Hall, December 15, 2008; Hyder, December 15, 2008.

21. "'Sit-Ins' Finally Hit Huntsville: Twenty-Three Jailed in 'Missile City,'" *Pittsburgh Courier*, January 20, 1962, 14.

22. "New Alabama Sit-Ins," *New York Times*, January 5, 1962, 18.

23. Ellis, "Space Crescent II," 240.

24. Anderson, "Sit-Ins," 14, col. 1.

25. Hereford and Ellis, *Beside the Troubled Waters*, 86.

26. Anderson, "Sit-Ins," 14, col. 1.

27. Ibid.

28. Dunar and Waring, *Power to Explore*, 116.

29. "Student Tells How CORE Staged Huntsville Sit-In," *Daily Defender*, January 31, 1962, 6, col. 1.

30. Dunar and Waring, *Power to Explore*, 117.

31. "Abduct, Spray White Alabama Integrationist," *Daily Defender*, January 23, 1962, 2, col. 4.

32. Testimony before the Committee of Inquiry into the Administration of Justice in the Freedom Struggle, May 25–26, 1962, http://content.cdlib.org/view?docId=kt800005x7&&doc.view=entire_text.

33. Hereford and Ellis, *Beside Troubled Waters*, 96, 89.

34. "Mrs. Roosevelt's Civic Unit to Eye Dixie Police Tactics," *Afro American*, May 26, 1962, 16.

35. Testimony before the Committee of Inquiry.

36. JML, "Report on Workshop for the Huntsville Movement," March 9, 1962, 1, http://www.thekingcenter.org/archive/document/report-workshop-huntsville-movement#.

37. Hereford and Ellis, *Beside Troubled Waters*, 90.

38. JML, "Report," 1.

39. Hereford and Ellis, *Beside Troubled Waters*, 94.

40. JML, "Report," 1.

41. Hereford and Ellis, *Beside Troubled Waters*, 101–103.

42. "Jail 6, Pregnant Women 90 Days for Sitting-In," *Daily Defender*, April 24, 1962, 5, col. 1.

43. Hereford and Ellis, *Beside Troubled Waters*, 104.

44. Ibid., 89.

45. Hall, December 15, 2008.

46. Hereford and Ellis, *Beside Troubled Waters*, 104–105.

47. Ibid., 105–106.

48. "No Dress-Up in Student Protest," *Daily Defender*, April 25, 1962, 15, col. 2.

49. Hereford and Ellis, *Beside Troubled Waters*, 106.

50. Marshall Frady, *Wallace: The Classic Portrait of Alabama Governor George Wallace* (New York: Random House, 1996), part 3.

51. "A Look Back to May 16–22, 1962," *Hartselle Enquirer*, http://www.hartselleenquirer.com/2012/05/16/a-look-back-to-may-16–22–1962/.

52. Stephan Lesher, *George Wallace: American Populist* (Menlo Park, CA: Addison Wesley, 1994), 157.

53. Hereford and Ellis, *Beside Troubled Waters*, 108.

54. JML, "Report," 4.

55. Alvin Spivak, Charlotte Moulton, William J. Eaton, Al Kuettner, and H. L. Stevenson, "Segregation: Is the Dam Breaking in Dixie, North?" *Chicago Defender* (national edition), June 8, 1963, 9.

56. Memorandum, Bart Slattery to Wernher von Braun, June 5, 1962, von Braun Papers, Library of Congress, box 8, 1962 D-E.

57. Ibid.

58. Hereford and Ellis, *Beside Troubled Waters*, 109.

59. Ibid.

60. Ibid., 100–111.

61. Ibid., 110.

62. James Miller, *Democracy Is in the Streets: From Port Huron to the Siege of Chicago* (Boston: Harvard University Press, 1994), 162; Simon Hal, "Protest Movements in the 1970s: The Long 1960s," *Journal of Contemporary History* 43, no. 4 (October 2008): 655.

63. Mary Ann Moore, oral history interview with Mary Moore, August 17, 2006, interview U-0193, Southern Oral History Program Collection (#4007), http://docsouth.unc.edu /sohp/U-0193/excerpts/excerpt_9219.html.

64. Memorandum, W. E. Guilian, chief counsel to V. C. Sorensen, chief, Management Services Office, February 26, 1963, "Edward Earl Morton Discrimination Case," MSFC Equal Opportunity Employment series, Digital Media, NASA Marshall Space Flight Center Historical Reference Collection, Huntsville, AL.

65. Jennings, November 20, 2008; Hyder, December 15, 2008; Hall, December 15, 2008.

66. Hereford and Ellis, *Beside Troubled Waters*, 26.

67. Hyder, December 15, 2008; Clyde Foster, conversation with Richard Paul, November 13, 2008; Hall, December 15, 2008.

68. Anthony Balderrama, "What Was Your Salary Worth Fifty Years Ago?" MSN Careers, http://msn.careerbuilder.com/Article/MSN-2146-Salaries-Promotions-What-was-your -salary-worth-50-years-ago/.

69. "Career Planning '64: A Special Supplement of the Afro American Newspapers," *Baltimore Afro-American*, March 21, 1964, 4, 5, 17, 32, 33, 62. A similar supplement was published March 27, 1965. R. Lynn Rittenoure, "Federal Employment in the Sixties," in *Employment of Blacks in the South: A Perspective on the 1960s*, ed. Ray Marshall and Virgil I. Christian (Austin: University of Texas Press, 1978), 147.

70. Report of the meeting held by representatives of the Huntsville contractors, 2, in Equal Opportunity Employment series, NASA Marshall Space Flight Center Historical Reference Collection, Huntsville, AL.

71. Letter, Wernher von Braun to James Webb, July 15, 1963, Equal Opportunity Employment series, Digital Media, NASA Marshall Space Flight Center Historical Reference Collection, Huntsville, AL.

72. Foster, November 13, 2008.

73. E. C. Smith, conversation with Richard Paul, December 15, 2008.

74. Foster, November 13, 2008.

75. Northrup, *Aerospace Industry*, 10.

76. Arthur Sanderson, personnel officer, Marshall Space Flight Center, NACA/NASA Equal Employment Opportunity Programs 1923–1992, file 188977, NASA EEO, e. 1960s Federal Records Center, NASA Headquarters Historical Reference Collection, Washington, DC, 3.

77. Hyder, December 15, 2008.

78. Hall, December 15, 2008.

79. Jennings, November 20, 2008.

80. Hall, December 15, 2008.

81. Wallace, "The 1963 Inaugural Address," 2, 9; Sitton, "Segregationists Fight," E5.

82. "Alabama Cancels Course after Negro Applies," *Atlanta Daily World*, February 2, 1963, 1, col. 7.

83. Ibid.

84. Lewis W. Jones, "Two Years of Desegregation in Alabama," *Journal of Negro Education* 25, no. 3 (Summer 1956): 206.

85. Ibid.

86. Ibid., 205, 206.

87. "Alabama Cancels," 1, col. 7.

88. E. Culpepper Clark, *The Schoolhouse Door: Segregation's Last Stand at the University of Alabama* (New York: Oxford University Press, 1993), 172.

89. Robert E. Baker, "Alabama Desegregation May Start at Huntsville," *Washington Post*, March 25, 1963, A2.

90. Ibid.

91. Ellis wrote of interviewing David McGlathery "as we sat in his car and talked late into a hot Alabama night," but in a January 2013 phone conversation with Richard Paul, McGlathery said any suggestion that Stuhlinger or anyone else convinced him not to enroll at UA-HC was not true. Ellis, "Space Crescent II," 240; David McGlathery, conversation with Richard Paul, January 17, 2013.

92. Clark, *Schoolhouse Door*, 236.

93. Baker, "Alabama Desegregation May Start," A2.

94. Clark, *Schoolhouse Door*, 172.

95. David McGlathery is disinclined to talk with outsiders anymore and would only answer "yes" or "no" in a brief conversation for this book.

96. Hall, December 15, 2008.

97. Ellis, "Space Crescent II," 240; memorandum, Burke Marshall to Robert F. Kennedy, April 9, 1963, "Civil Rights: Alabama, University of 4/9/63–5/31/63" folder, Papers of Robert F. Kennedy, Attorney General's General Correspondence, box 10, JFKL.

98. Al Kuettner, "How Courageous Businessmen Saved Day in Birmingham, South's Racial Powder Keg," *Chicago Defender*, May 18, 1963, 19.

99. Martin Luther King Jr., "Letter from a Birmingham Jail," University of Pennsylvania African Studies Center, http://www.africa.upenn.edu/Articles_Gen/Letter_Birmingham.html.

100. Letter, James E. Webb to Wernher von Braun and Kurt H. Debus, April 19, 1963, Minority Groups (1961–1993), file 008983, NASA Headquarters Historical Reference Collection, Washington, DC.

101. "U.S. Says It Can't Act to Halt Birmingham Bias," *Chicago Daily Defender*, April 15, 1963, 4.

102. Letter, James E. Webb to Wernher von Braun and Kurt H. Debus, 19 April 1963, James E. Webb Files, NASA Historical Reference Collection, NASA Headquarters, Washington, DC.

103. Launius, "Managing the Unmanageable," 159.

104. James Webb in "Conference on Space, Science, and Urban Life: Proceedings of a Conference Held at Oakland, California," March 28–30, 1963, National Aeronautics and Space Administration, 94 (Washington, DC: Office of Scientific and Technological Information, National Aeronautics and Space Administration, 1963).

105. Launius, "Managing the Unmanageable," 162.

106. Telephone call, Alfred S. Hodgson to von Braun staff, May 23, 1963, 9:30 a.m., NARA Southeast Region, Record Group 255, Accession number 01–0002, upper-level management files, box 23.

107. Ibid.

108. Clark, *Schoolhouse Door*, 195.

109. Memorandum, William H. Orrick Jr. to Robert F. Kennedy, May 29, 1963, "Civil Rights: Alabama, University of 4/9/63–5/31/63" folder, Papers of Robert F. Kennedy, Attorney General's General Correspondence, box 10, JFKL.

110. Memorandum, James E. Webb to Alfred Hodgson, May 29, 1963, "May 1963" folder, chronological file, James E. Webb Personal Papers, JFKL.

111. Letter, J. A. Barclay to George C. Wallace, May 23, 1963, Alabama Department of Archives and History Digital Collections, Alabama Textual Materials Collection, http://digital.archives.alabama.gov/cdm/singleitem/collection/voices/id/3900/rec/5.

112. Daily journal of Dr. von Braun, May 22, 1963, NASA Headquarters Historical Reference Collection, NASA Headquarters, Washington, DC.

113. Letter, Wernher von Braun to Fortune Ryan, June 27, 1963, Chief of Public Affairs, box 14, "Dr. von Braun—Personal," NARA Atlanta, RG255, MSFC/ULMF #70A1658.

114. Memorandum, Jack Rosenthal to Pierre Salinger, May 13, 1963, "Alabama" folder, box 95, Papers of Robert F. Kennedy, JFKL.

115. Clark, *Schoolhouse Door*, 210.

116. Ibid., 208.

117. Stephan Lesher, *George Wallace: American Populist* (Menlo Park, CA: Addison-Wesley, 1994), 233.

118. William O. Bryant, "David Mack McGlathery Admitted to Alabama U: Few Spectators on Hand as Race Barrier Falls," *Atlanta Daily World*, June 14, 1963, 1, col. 4.

119. Memorandum, Jack Rosenthal to Pierre Salinger, May 13, 1963, "Alabama" folder, box 95, Papers of Robert F. Kennedy, JFKL; memorandum, Burke Marshall to Robert F. Kennedy, May 22, 1963, "Civil Rights: Alabama, University of 4/9/63–5/31/63" folder, Papers of Robert F. Kennedy, Attorney General's General Correspondence, box 10, JFKL; Lesher, *George Wallace*, 216, 241; "Set to Go to U. of Ala.," *Baltimore Afro-American*, June 1, 1963, 1.

120. Hall, December 15, 2008.

121. Hyder, December 15, 2008.

122. Graham, *Civil Rights Era*, 79.

Chapter 8

1. James H. Peyton, "Michigan News," *Chicago Defender*, September 5, 1925, A6.

2. Hamilton Bims, "Rocket Age Comes to Tiny Triana: Space Technologist Revives Forgotten Alabama Town," *Ebony*, March 1965, 111; Sam Fulwood, "Today's Big-City Black Mayors Lead in a Changed Political World," *Houston Chronicle*, September 24, 1995, sec. A, 22.

3. Foster, November 13, 2008.

4. Ibid.

5. Ibid.

6. There are numerous documents in the MSFC archives, many including von Braun's handwritten notations, that demonstrate this assertion.

7. Bims, "Rocket Age," 112. Bims' *Ebony* article highlighted Foster's work.

8. Foster, November 13, 2008.

9. "For Negroes: More and Better Jobs in Government," *U.S. News and World Report*, March 5, 1962, 83–84; U.S. Civil Service Commission, *Study of Minority Group Employment in the Federal Government, 1965* (Washington, DC: Government Printing Office, 1965), 25; U.S. Civil Service Commission, *Study of Minority Group Employment in the Federal Government, November 30, 1969* (Washington, DC: GPO, 1969), 310, 311; U.S. Commission on Civil Rights, *Federal Civil Rights Enforcement Effort* (Washington, DC: Government Printing Office, 1970), 68.

10. Foster, November 13, 2008; Watson, January 27, 2009; Clyde Foster, conversation with Richard Paul, January 11, 2010.

11. Foster, November 13, 2008.

12. John R. Seeley, Bertram M. Gross, Sumner Myers, Lewis A. Dexter, and Edward E. Furash, *Space, Society and Social Science*, paper presented at the Committee on Space Efforts and Society of the American Academy of Arts and Sciences, Cambridge, MA, February 1963, 6, 11, 13.

13. Bauer et al., *Second-Order Consequences*, 67.

14. Dodd, "Slow Pace of Change," 568.

15. Ibid., 569.

16. "Peter C. Dodd '50," *Princeton Alumni Weekly* 111, no. 10, April 6, 2011, http://paw .princeton.edu/issues/2011/04/06/sections/memorials/1960/index.xml.

17. Carruthers, June 15, 2009.

18. Dodd, "Slow Pace of Change," 573.

19. Ibid.

20. Ibid., 575.

21. Ibid., 576.

22. Ibid., 573.

23. Ibid.

24. Ibid., 574.

25. Button, *Blacks and Social Change*, 72, 185.

26. Akens, *Rocket City, USA*, 59.

27. Dodd, "Slow Pace of Change," 574.

28. Bauer et al., *Second-Order Consequences*, 97.

29. Mary A. Holman, *The Political Economy of the Space Program* (Palo Alto, CA: Pacific Books, 1974), 200, 203, 205, 207.

30. Ibid., 206. Holman's employment figure included military personnel.

31. Bauer et al., *Second-Order Consequences*, 98–100.

32. Button, *Blacks and Social Change*, 71.

33. Ibid.; Bauer et al., *Second-Order Consequences*, 96–101.

34. Bauer et al., *Second-Order Consequences*, 75, 98.

35. Foster, November 13, 2008.

36. Memorandum, Guilian to Sorensen, February 26, 1963.

37. Dunar and Waring, *Power to Explore*, 118.

38. Letter, James E. Webb to Wernher von Braun and Kurt H. Debus, September 9, 1963, Equal Opportunity Employment series, Digital Media, NASA Marshall Space Flight Center Historical Reference Collection, Huntsville, AL.

39. Letter, Harry H. Gorman to Alfred S. Hodgson, September 17, 1963, Equal Opportunity Employment series, Digital Media, NASA Marshall Space Flight Center Historical Reference Collection, Huntsville, AL.

40. Letter, Wernher von Braun to James E. Webb, September 30, 1963, Equal Opportunity Employment series, Digital Media, NASA Marshall Space Flight Center Historical Reference Collection, Huntsville, AL.

41. Letter, Alfred S. Hodgson to Wernher von Braun, October 10, 1963, Equal Opportunity Employment series, Digital Media, NASA Marshall Space Flight Center Historical Reference Collection, Huntsville, AL.

42. Foster, November 13, 2008.

43. Ibid.

44. "Bombs Upset Racial Peace in Birmingham: Blast Home of M. L. King's Kin, Gaston Motel," *Chicago Defender*, May 11, 1963, 1.

45. Quoted in Arthur M. Schlesinger Jr., *Robert Kennedy and His Times* (Boston: Houghton Mifflin, 1978), 332.

46. Ibid.; Hilty, *Robert Kennedy*, 358; Graham, *Civil Rights Era*, 71.

47. Graham, *Civil Rights Era*, 67; Schlesinger, *Robert Kennedy*, 336, 337.

48. Allen Fisher (archivist, LBJ Library), e-mail correspondence with Richard Paul, January, 8, 2013.

49. Memorandum, Young to Kier, June 6, 1963.

50. Letter, James E. Webb to James B. Morrison, president, Chesapeake and Potomac Telephone Company, June 6, 1963, James E. Webb Files, NASA Historical Reference Collection, NASA Headquarters, Washington, DC.

51. Memoranda, James E. Webb to Robert Seamans, May 20, 1963; James E. Webb to Alfred Hodgson, May 24, 1963, "May 1963" folder, chronological file; James E. Webb to Paul Dembling and Richard Callaghan, August 16, 1963, "August 1963" folder, chronological file, James E. Webb Personal Papers, JFKL.

52. Congress, Senate, Committee on Aeronautical and Space Sciences, *NASA Authorization for Fiscal Year 1964, pt. 2: Program Detail*, 88th Cong., 1st sess., 12, 13, 17, June 18, 1963, 903.

53. Watson, January 27, 2009.

54. Congress, Senate, Senator Javits of New York speaking on discriminatory use of federal funds, 88th Cong., 2nd sess., *Congressional Record* (January 30, 1964), 110, pt. 2, 1391–1392.

55. U.S. President, *Public Papers of the Presidents of the United States* (Washington, DC: Office of the Federal Register, National Archives and Records Service, 1962), John F. Kennedy, 1961, 304.

56. Letter, Mack Herring, chief, Public Affairs Office, NACA/NASA Equal Employment Opportunity Programs 1923–1992, file 188977, NASA EEO, e. 1960s Federal Records Center, NASA Headquarters Historical Reference Collection, Washington, DC.

57. Letter, James E. Webb to Wernher von Braun, June 24, 1963, Minority Groups (1961–1993), file 008983, NASA Headquarters Historical Reference Collection, Washington, DC.

58. The letter specified that the contractors who needed to attend were Management Services, SPACO, Brown Engineering, General Electric, Hayes International, White Castle, Federal Services, Wetland, Technical Productions, and W. T. Schrimisher.

59. Webb to von Braun, June 24, 1963.

60. Michael J. Neufeld, *Von Braun: Dreamer of Space, Engineer of War* (New York: Alfred A. Knopf, 2007), 397.

61. Bob Ward, *Dr. Space: The Life of Wernher von Braun* (Annapolis, MD: Naval Institute Press, 2005), 173.

62. Report of the meeting held by representatives of the Huntsville Contractors, MSFC Division Files (Atlanta), July 5, 1963, 1, box 1, Equal Opportunity Employment series, Digital Media, NASA Marshall Space Flight Center Historical Reference Collection, Huntsville, AL.

63. Memorandum, Hobart Taylor Jr. to Charles A. Horsky, July 29, 1963, "Coordinating Program for Equal Opportunity Housing for Federal Employees," office files of Lee C. White, box 3, LBJL.

64. Report of the meeting held by representatives of the Huntsville Contractors, 3, 1.

65. "Separate racial station" comes from Wallace's first inaugural address, George C. Wallace, "The 1963 Inaugural Address," p. 2; Report of the meeting held by representatives of the Huntsville Contractors, 2–3.

66. Letter, Wernher von Braun to James Webb, July 15, 1963, Equal Opportunity Employment series, Digital Media, NASA Marshall Space Flight Center Historical Reference Collection, Huntsville, AL.

67. William S. Ellis, "Space Crescent II," 241; Congress, Senate, Senator Javits of New York speaking on discriminatory use of federal funds, 88th Cong., 2nd sess., *Congressional Record* (January 30, 1964), 110, pt. 2, 1391–1393. Senator Javits included Paul G. Dembling's letter in his remarks to the Senate.

68. "Status of Desegregation of Thirty-Nine Selected Southern Cities: March 1964," 3/64, "Civil Rights-Ad Hoc Businessmen's Group" folder, box 4, office files of Lee C. White, LBJL.

69. In Arthur Schlesinger's *Robert Kennedy and His Times*, which contains the seminal description of this confrontation, Conway is Schlesinger's source for the PCEEO story. However, the July 18 transcript does not list him as a participant. Allen Fisher, an archivist at the LBJ Library, said that, traditionally, Conway would only come to meetings when Walter Reuther of the United Auto Workers could not attend. Reuther was at the July 18 PCEEO meeting. However, the July 18 transcript includes speech from people who are not listed as attending, so there is a chance Conway was there but was simply not listed as attending. Schlesinger, *Robert Kennedy*, 36.

70. Memorandum, Hobart Taylor Jr. to Lyndon B. Johnson, July 12, 1963, James E. Webb Files, NASA Historical Reference Collection, NASA Headquarters, Washington, DC.

71. President's Committee on Equal Employment Opportunity, July 18, 1963, LBJL, 17–18.

72. Ibid., 31.

73. Ibid.

74. Schlesinger, *Robert Kennedy*, 336.

75. President's Committee on Equal Employment Opportunity, 34–35.

76. Ibid., 37–38.

77. Schlesinger, *Robert Kennedy*, 336.

78. In *Robert Kennedy and His Times,* Schlesinger wrote that Robert Kennedy expressed his general fury at the poor state of equal employment opportunity during a PCEEO meeting on May 29 and then became angry at James Webb specifically during another PCEEO meeting "twenty days later," which would have been June 18. However, the confrontation Schlesinger recounts, which he quotes verbatim, appears in the July 18 PCEEO transcript. The June 18 mistake is repeated in *Power to Explore,* the MSFC official history. In his Robert Kennedy book *Brother Protector,* James Hilty suggests Kennedy threatened to fire Webb at the May 29 meeting, which Webb did not attend. In a 2006 book on Kennedy and civil rights called *The Bystander,* Nick Bryant repeated the May 29 date, saying that RFK "unleashed a barrage of barbed questions," attacked Webb, screamed at Hobart Taylor, and then "stormed out."

79. James W. Button said in *Blacks and Social Change,* "There is still widespread debate about the effects of the civil rights movement" on African Americans. Harvard Sitkoff, Allan J. Matusow, and others have argued that the civil rights reforms of the 1960s were a failure because they addressed only political power and not economic opportunity; this opinion is summed up by Hugh Davis Graham: "social and economic equality were pursued at a procedural level that, like the War On Poverty, left substantive structures of inequality intact"; Graham, *Civil Rights Era,* 451–452.

80. Foster, November 13, 2008, January 11, 2010.

81. Jennings, November 20, 2008; James Jennings, conversation with Allan Needell, July 23, 2010, http://capecosmos.org.

82. Jennings, November 20, 2008.

83. Foster, November 13, 2008.

84. Jennings, July 23, 2010.

85. Watson, January 27, 2009.

86. Bims, "Rocket Age," 106.

87. Ibid.

88. Foster, November 13, 2008.

89. Dorothy Foster, conversation with Richard Paul, July 19, 2012.

90. Foster, November 13, 2008.

91. Bims, "Rocket Age," 108, 112.

92. Ibid., 108.

93. Foster, November 13, 2008.

94. Bims, "Rocket Age," 111.

95. Ibid.

96. Foster, November 13, 2008.

97. Ibid.

98. Ibid.

99. Foster, January 11, 2010.

Chapter 9

1. Watson, January 27, 2009. Walter Applewhite and Wesley Carter worked at the Redstone Arsenal's Army Missile Command in Huntsville rather than at NASA.

2. Ibid.

3. Morgan Watson, panel discussion, National Air and Space Museum, February 20, 2010.

4. Watson, January 27, 2009.

5. George Bourda, conversation with Richard Paul, November 15, 2012.

6. Watson, January 27, 2009.

7. Editorial, "Southern University Plantation," *Chicago Defender*, February 24, 1962, 8.

8. "Thirty-Five More Sit-In Students Jailed," *Daily Defender*, March 30, 1960, 3; "US Supreme Court Affirms N.O. School Desegregation," *Pittsburgh Courier*, April 1, 1961, A44; UPI, "La. Sit-In Convictions Overturned," *Chicago Defender*, December 12, 1961, 1.

9. Bourda, November 15, 2012.

10. "Southern University Plantation," 8; Robert Cohen and David J. Snyder, eds., *Rebellion in Black and White: Southern Student Activism in the 1960s* (Baltimore: Johns Hopkins University Press, 2013), 4.

11. Bourda, November 15, 2012.

12. Watson, January 27, 2009.

13. Michele M. Simms Paris, "What Does It Mean to See a Black Church Burning? Understanding the Significance of Constitutionalizing Hate Speech," *University of Pennsylvania Journal of Constitutional Law* (Spring 1998): 138; Kraig Beyerlein and Kenneth T. Andrews, "Black Voting during the Civil Rights Movement: A Micro-Level Analysis," *Social Forces* 87, no. 1 (September 2008): 8; Glenn T. Eskew, *But for Birmingham: The Local and National Movements in the Civil Rights Struggle* (Chapel Hill: University of North Carolina Press, 1997), 127–128, 323; Cohen and Snyder, *Rebellion in Black and White*, 4.

14. Charles Smoot, conversation with Richard Paul, January 28, 2013.

15. Finney, "NASA Is Training," 54; Hall, December 15, 2008; MSFC Manpower Office, "A Chronology of the Equal Employment Opportunity Program at MSFC," February 1971, Equal Opportunity Employment series, Digital Media, NASA Marshall Space Flight Center Historical Reference Collection, Huntsville, AL, 6.

16. Earnest C. Smith, conversation with Richard Paul, December 15, 2008.

17. Smoot, January 28, 2013.

18. Smith, December 15, 2008.

19. Smoot, January 28, 2013.

20. Letter, Wernher von Braun to James Webb, July 15, 1963, Equal Opportunity Employment series, Digital Media, NASA Marshall Space Flight Center Historical Reference Collection, Huntsville, AL.

21. Smoot, January 28, 2013.

22. Bourda, November 15, 2012.

23. Smoot, January 28, 2013.

24. Betteridge's "law of headlines" is a twenty-first-century adage declaring that "any headline which ends in a question mark can be answered by the word no"; the opposite is often true, however, with the question mark serving as a fig leaf to maintain journalistic objectivity. In fact, as Roger Simon wrote, "In journalism, a question mark (in a headline) justifies virtually anything, no matter how unlikely." Roger Simon, "Empty Seats Haunt President Obama," *Politico*, May 8, 2012, http://www.politico.com/news/stories/0512/76017.html.

25. "Negro College Youth to Boost First Moongoer into Orbit?" *Chicago Defender*, March 14, 1964, 3.

26. In a 1964 article, the *New York Times* said there were eleven African Americans, "about half" professional and the rest clerical. But it is difficult to know which facility the reporter was writing about because he conflated anecdotes and statistics from Florida and Alabama. The dateline on the story was Washington, DC.

27. Finney, "NASA Is Training," 54.

28. Watson, January 27, 2009.

29. Report of the meeting held by representatives of the Huntsville Contractors, MSFC Division Files (Atlanta), 1963, box 1, Equal Opportunity Employment series, Digital Media, NASA Marshall Space Flight Center Historical Reference Collection, Huntsville, AL, July 5, 1963, 3.

30. Watson, January 27, 2009.

31. Ibid.

32. Bourda, November 15, 2012.

33. Smoot, January 28, 2013.

34. Watson, January 27, 2009.

35. Bourda, November 15, 2012.

36. Wallace, "The 1963 Inaugural Address," p. 2.

37. Watson, January 27, 2009.

38. Bourda, November 15, 2012.

39. Watson, January 27, 2009.

40. Bourda, November 15, 2012.

41. Watson, January 27, 2009.

42. Ray, February 16, 2009.

43. The details and the dialogue from this anecdote come from interviews with George Bourda and Morgan Watson.

44. Finney, "NASA Is Training," 54.

45. Morgan Watson, conversation with Thomas Lassman, July 23, 2010, http://cape cosmos.org.

46. Bourda, November 15, 2012.

47. Watson, January 27, 2009.

48. Bourda, November 15, 2012.

49. Amy E. Foster, *Integrating Women into the Astronaut Corps: Politics and Logistics at NASA, 1972–2004* (Baltimore: Johns Hopkins University Press, 2011), 101.

50. Watson, January 27, 2009.

51. Bourda, November 15, 2012.

52. Watson, January 27, 2009.

53. Robert A. Caro, conversation with Bat Segundo, May 15, 2012, http://www.edrants .com/the-bat-segundo-show-robert-a-caro/.

54. Telephone conversation, President Johnson to Walter Jenkins, January 28, 1964, tape 1530, WH 6401.20 Program Number 31, LBJL.

55. Tom Wicker, "Remembering the Johnson Treatment," *New York Times*, May 9, 2002, http://www.nytimes.com/2002/05/09/opinion/remembering-the-johnson-treatment.html.

56. Telephone conversation, President Johnson to John Stennis, January 28, 1964, tape 1594, WH 6401.23 PNO 13, LBJL.

57. "Urge Halt to NASA Officials Speech Jackson MS.," *Informer (Houston)*, February 1,

1964, 2. President Johnson's "rule" is a reference to Executive Order 10925 and President Kennedy's "Memorandum on Racial or Other Discrimination in Federal Employee Recreational Associations."

58. "Party Is Boycotted by the Astronauts," *New York Times*, March 3, 1965, 33; "Party Snubbed by Astronauts," *Informer (Houston)*, March 6, 1965, 1.

59. Letter, Harry H. Gorman to Alfred S. Hodgson, March 14, 1964, Equal Opportunity Employment series, Digital Media, NASA Marshall Space Flight Center Historical Reference Collection, Huntsville, AL.

60. Von Braun daily journal, September 4, 1964, NASA Headquarters, Historical Reference Collection, Washington, DC, 13259.

61. Graham, *Civil Rights Era*, 75.

62. Letter, James Webb to Hon. Robert R. Casey, November 17, 1964, Webb Oct. 1964 Huntsville, AL, Controversy, file 3517, NASA Headquarters Historical Reference Collection, Washington, DC.

63. John W. Finney, "NASA May Leave Its Alabama Base," *New York Times*, October 24, 1964, 12.

64. Wernher von Braun, "Speech to the Alabama Legislature on June 8," June 8, 1965, NASA Headquarters Historical Reference Collection, Washington, DC, Speeches by Key Officials, Wernher von Braun, https://mira.hq.nasa.gov/history/.

65. Telegram, Jim Dunn to President Johnson, October 27, 1964, 4/15/64–10/31/64, FG 260, box 295, LBJL.

66. James E. Webb, "Discussion of George C. Marshall Space Flight Center Programs and Problems, October 29, 1964," Webb Oct 1964 Huntsville, AL, Controversy, file 3517, NASA Headquarters Historical Reference Collection, Washington, DC; NASA Historical Staff, *Astronautics and Aeronautics, 1964: Chronology on Science, Technology, and Policy* (Washington, DC: Government Printing Office, 1965), 367.

67. UPI, "Ala. Bigotry May Cause Space Research Loss," *Los Angeles Sentinel*, November 5, 1964, A11; William Hines, "Point of View," *Washington Star*, October 26, 1964; Webb Oct. 1964 Huntsville, AL, Controversy, file 3517, NASA Headquarters Historical Reference Collection, Washington, DC.

68. Neufeld, *Von Braun*, 396.

69. UPI, "NASA Refuses to Reveal Names," Webb Oct. 1964 Huntsville, AL, Controversy, file 3517, NASA Headquarters Historical Reference Collection, Washington, DC.

70. Neufeld, *Von Braun*, 395.

71. Letter, J. A. Barclay to George C. Wallace, May 23, 1963, Alabama Department of Archives and History Digital Collections, Alabama Textual Materials Collection, http://digital.archives.alabama.gov/cdm/singleitem/collection/voices/id/3900/rec/5.

72. William S. Ellis, "Space Crescent III: The Wide Blue Porkbarrel," *Nation*, October 26, 1964, 276.

73. Clay Risen, "Suburbs, not Racism, Blamed for Political Shift in South," *Atlanta Journal-Constitution*, April 30, 2006, 5K.

74. *Aviation Week and Space Technology*, November 2, 1964, 15; Webb Oct. 1964 Huntsville, AL, Controversy, file 3517, NASA Headquarters Historical Reference Collection, Washington, DC.

75. Finney, "NASA May Leave," 12.

76. "NASA Job Talk Scored by Burch as 'Browbeating,'" *Sunday Star*, November 1, 1964; Webb Oct. 1964 Huntsville, AL, Controversy, file 3517, NASA Headquarters Historical Reference Collection, Washington, DC.

77. "Webb Repeats Warning to Marshall (an Analysis)," *Space Business Daily*, November 2, 1964, 1–2, Webb Oct. 1964 Huntsville, AL, Controversy, file 3517, NASA Headquarters Historical Reference Collection, Washington, DC.

78. "1964 Presidential General Election Results," http://uselectionatlas.org/RESULTS/national.php?year=1964.

79. Neufeld, *Von Braun*, 392.

80. "One Hundred of Space Agency Will Leave Alabama," *New York Times*, November 28, 1964, 16.

81. Letter, Wernher von Braun to Bert Slattery, November 10, 1964, Wernher von Braun Papers, U.S. Space and Rocket Center History Office, Huntsville, AL, 123/12.

82. Wernher von Braun, speech, annual banquet—Huntsville/Madison County Chamber of Commerce—December 8, 1964, 13, Speeches by Key Officials, Wernher von Braun, NASA Headquarters Historical Reference Collection, https://mira.hq.nasa.gov/history/.

83. Ibid., 9, 15.

84. Neufeld, *Von Braun*, 396.

85. Wernher von Braun, "Building a Space Program" (speech), State Home Builders Association, Mobile, AL, September 28, 1964, Speeches by Key Officials, Wernher von Braun, NASA Headquarters Historical Reference Collection, https://mira.hq.nasa.gov/history/.

86. Jonathan McPherson, organizer of the Miles College Student Boycott of Segregated Birmingham Businesses, "First Person Accounts from 1963 Birmingham Campaign," Birmingham Civil Rights Institute, April 25, 2013, http://www.c-span.org/video/?312240-1/first-person-accounts-1963-birmingham-campaign. http://www.c-spanvideo.org/program/FirstPe.

87. Neufeld, *Von Braun*, 396; "Miles College Gets Science Building, $50,000," *Jet Magazine*, December 10, 1964, 23; "History Began Again," *Chicago Daily Defender*, December 9, 1964, 10.

88. Letter, L. C. McMillan to Wernher von Braun, November 30, 1961–64, Equal Opportunity Employment series, Digital Media, NASA Marshall Space Flight Center Historical Reference Collection, Huntsville, AL.

89. Letter, Linton Crook to Wernher von Braun, April 19, 1965, reproduced at National Archives and Records Administration.

90. Lyndon B. Johnson, *The Vantage Point: Perspectives of the Presidency, 1963–1969* (Chicago: Holt, Rinehart and Winston, 1971), 161–166; Lesher, *George Wallace*, 323–334.

91. Robert E. Baker, "Wallace Clouds the Alabama Murk: Visiting Newspapermen Were Treated Royally but Kept from the Gut Issue," *Washington Post*, June 13, 1965, E1.

92. University of Alabama Board of Trustees, "A Chronological History of the University of Alabama at Birmingham (UAB) and Its Predecessor Institutions and Organizations, 1831–," http://www.uab.edu/archives/chron.

93. Memorandum, James E. Webb to Dr. Simpson, May 28, 1965, box 1, AC 69–88, James E. Webb Personal Papers, LBJL.

94. Telephone conversation, Werhner von Braun with James Webb, May 26, 1965, 13260, WvBP-4 122–1, NASA Headquarters, Historical Records Collection, Washington, DC.

95. Telephone conversation, Wernher von Braun with John Zierdt, May 26, 1965, 13260, WvBP-4 122–1, NASA Headquarters, Historical Records Collection, Washington, DC.

96. Telephone conversation, von Braun with Webb, May 26, 1965.

97. Ibid.

98. Telephone conversation, Wernher von Braun with Col. Lawrence W. Vogel, May 26, 1965, 13260, WvBP-4 122–1, 5, NASA Headquarters, Historical Records Collection, Washington, DC.

99. Baker, "Wallace Clouds," E1.

100. Von Braun, "Speech to the Alabama Legislature on June 8," June 8, 1965.

101. Ben A. Franklin, "Wallace Is Given a NASA Warning," *New York Times*, June 9, 1965, 31.

102. Cox News Service, "Voice from Vanguard of Civil Rights Movement James Farmer, Last of a Generation of Black Leaders, Talks of Hope," *Baltimore Sun*, May 15, 1997, http://articles.baltimoresun.com/1997–05–15/news/1997135088_1_rights-movement-james-farmer-civil-rights.

103. Whitney M. Young, "Third of Moon Trip Cost Could End Poverty," *Rock Hill Herald*, August 2, 1969, http://news.google.com/newspapers?nid=1821&dat=19690802&id=eUYtAAAAIBAJ&sjid=b58FAAAAIBAJ&pg=4883,2772269.

104. A master's thesis entitled "Space Race: African American Newspapers Respond to *Sputnik* and Apollo 11," written in 2007 by Mark A. Thompson of the University of North Texas, looked at articles from this period in the *New York Amsterdam News*, *Los Angeles Sentinel*, *Michigan Chronicle*, *Oklahoma City Black Dispatch*, *Milwaukee Courier*, and *Minneapolis Spokesman* and found they uniformly called into question the amount of money spent on space exploration, comparing it with the needs of the hungry and poverty-stricken. Many of these writers invoked Abernathy's protest march.

105. James Jennings, conversation with Allan Needell, July 23, 2010, http://capecosmos.org.

106. Bourda, November 15, 2012; Watson, January 27, 2009.

107. Watson, January 27, 2009.

Conclusion

1. Jennings, November 20, 2008.

2. Nanette Dobrosky, *A Guide to the Microform Edition of the Presidential Oral History Series: The John F. Kennedy Presidential Oral History Collection* (Frederick, MD: University Publications of America, 1988), v–vii.

3. Hobart Taylor Jr., oral history interview by John F. Stewart, January 11, 1967, 23, 24, JFKL.

4. Watson, January 27, 2009.

5. Arthur Sanderson, personnel officer, *Marshall Space Flight Center*, NACA/NASA Equal Employment Opportunity Programs 1923–1992, file 188977, NASA EEO, e. 1960s Federal Records Center, NASA Headquarters Historical Reference Collection, Washington, DC.

6. Kim McQuaid, "Racism, Sexism, and Space Ventures: Civil Rights at NASA in the Nixon Era and Beyond," in *Societal Impact of Spaceflight*, ed. Steven J. Dick and Roger D. Launius (Washington, DC: NASA History Division, 2007), 421; Dunar and Waring, *Power*

to Explore, 124; "NASA Equal Opportunity Program," appended to Ruth Bates Harris, Joseph M. Hogan, and Samuel Lynn to James C. Fletcher, September 20, 1973, NASA History Division Documents Collection, NASA Headquarters, Washington, DC.

7. Smoot, January 28, 2013.

8. Delano Hyder, Richard Hall, and E. C. Smith, conversation with Richard Paul, December 15, 2008.

9. Button, *Blacks and Social Change*, 71.

10. Ibid.

11. Ibid., 71, 73; Bauer et al., *Second-Order Consequences*, 96–101; Annie Mary Hartsfield, Mary Alice Griffin, and Charles M. Grigg, eds., *Summary Report: NASA Impact on Brevard County* (Tallahassee, FL: Institute for Social Research, 1966), 16–19.

12. Susanne Cervenka, "Melbourne City Council Vote May End Diversity Drought: City Hasn't Seen Black Council Member since 1977 Elections," *Florida Today*, November 3, 2012, http://www.floridatoday.com/article/20121103/NEWS05/311030019/Melbourne-city -council-vote-may-end-diversity-drought.

13. "For Negroes: More and Better Jobs in Government," *U.S. News and World Report*, March 5, 1962, 83–84; U.S. Civil Service Commission, *Study of Minority Group Employment in the Federal Government, 1965* (Washington, DC: Government Printing Office, 1965), 25; U.S. Civil Service Commission, *Study of Minority Group Employment in the Federal Government, November 30, 1969* (Washington, DC: Government Printing Office, 1969), 310, 311; U.S. Commission on Civil Rights, *Federal Civil Rights Enforcement Effort* (Washington, DC: Government Printing Office, 1970), 68.

14. Tables A.1 to A.8 in the appendix contain relevant census numbers for the NASA host states of Alabama, Florida, Mississippi, and Texas.

15. U.S. Department of Commerce, *1960 Census of the Population, Vol. 1: Characteristics of the Population, Pt. 11: Florida* (Washington, DC: Government Printing Office, 1963), 11-469, table 122.

16. U.S. Department of Commerce, *1960 Census of the Population, Vol. 1: Characteristics of the Population, Pt. 2: Alabama* (Washington, DC: Government Printing Office, 1963), 2-373, 379, tables 121 and 122; U.S. Department of Commerce, *1970 Census of the Population, Vol. 1: Characteristics of the Population, Pt. 2: Alabama* (Washington, DC: Government Printing Office, 1973), 2-623, 626, table 173.

17. Letter, Robert E. Jones to President Johnson, January 23, 1967, General FG 260 8/12/65, FG 260, box 295, LBJL; memorandum, James E. Webb to Richard Callaghan, March 14, 1967, March 1967, box 2, James E. Webb Personal Papers, LBJL. There appears to be no mention of the Alabama race question after 1967 in the James E. Webb Personal Papers at the LBJ Library.

18. U.S. Department of Commerce, *1960 Census of the Population, Vol. 1: Characteristics of the Population, Pt. 45: Texas* (Washington, DC: Government Printing Office, 1963), 45-884, table 122; U.S. Department of Commerce, *1970 Census of the Population, Vol. 1: Characteristics of the Population, Pt. 45: Texas, Section 2* (Washington, DC: GPO, 1973), 45-1679, 45-1683, table 173.

19. Northrup, *Aerospace Industry*, 22.

20. Otis King, conversation with Richard Paul, December 24, 2008.

21. "The Space Age Comes to Mississippi," 80; letter, James E. Webb to Adam Clayton Powell, February 4, 1964, General FG 260 NASA 11/22/63-2/24/64, WHCF, box 294, LBJL.

22. U.S. Department of Commerce, *1960 Census of the Population, Vol. 1: Characteristics of the Population, Pt. 26: Mississippi* (Washington, DC: Government Printing Office, 1963), 26-333, table 122.

23. U.S. Department of Commerce, *1970 Census of the Population, Vol. 1: Characteristics of the Population, Pt. 26: Mississippi* (Washington, DC: Government Printing Office, 1973), 26-553, 535, table 173.

24. Herbert R. Northrup, "In-Plant Movement of Negroes in the Aerospace Industry," *Monthly Labor Review* 91 (February 1968): 23.

25. Northrup, *Aerospace Industry*, 10.

26. Ibid.

27. Jennings, November 20, 2008.

28. "Excerpts of Remarks by Dr. George M. Low before Conference on EEO," undated (apparently 1973), NASA History Division Documents Collection, NASA Headquarters, Washington, DC.

29. McQuaid, "Racism, Sexism, and Space Ventures," 431, 447.

30. "Race Labor Leaving," *Chicago Defender*, February 5, 1916, 1.

31. Isabel Wilkerson, remarks at "From Emancipation to the Great Migration," a National Constitution Center and New York Public Radio event at the Greene Space in New York City, January 21, 2013, http://www.c-span.org/Events/From-Emancipation-to-the -Great-Migration/10737437314/.

32. Ibid.

33. Watson, January 27, 2009.

34. Jennings, November 20, 2008.

35. Watson, January 27, 2009.

36. Wilkerson, "From Emancipation."

37. Carruthers, June 15, 2009.

38. Hyder, Hall, and Smith, December 15, 2009.

39. Watson, January 27, 2009.

40. Finney, "NASA Is Training," 54.

41. Watson, January 27, 2009.

42. Mae Jemison, conversation with Richard Paul, March 30, 2009.

43. Richard Paul arranged for Julius Montgomery and Morgan Watson to be at the National Air and Space Museum that day, moderated the panel on which they spoke, and witnessed this exchanged in the Green Room.

44. Smoot, January 28, 2013.

45. John F. Kennedy, "Moon Speech" (speech, Rice Stadium, Rice University, Houston, TX, September 12, 1962), http://er.jsc.nasa.gov/seh/ricetalk.htm.

Acknowledgments

The path that led to the men profiled in this book was long and took many turns. Among those who helped most in getting us to these remarkable characters are Raymond Archer, Bill Bell, Hattie Carwell, Kristen Clark, Tom Cole, Tremayne Dillard, Kimberly Dickerson, Steven Fischler, Roz Foster, Herschell Hamilton, Kerry-Ann Hamilton, Wes Harris, Jay Holloway, Andrea Hogans, Robert Mosby, Jim McWilliams, Gordon Patterson, Julieanna Richardson, Brent Ross, and Raimond Struble.

This manuscript was made possible in part by numerous grants that funded our research. Without the John F. Kennedy Library Foundation and the Lyndon B. Johnson Foundation, none of Steven's early work on NASA and racial equality would have been done. Generous funding from the National Science Foundation enabled Richard to find and record interviews with the African American pioneers whose stories are told in these chapters, while the Smithsonian National Air and Space Museum's (NASM) Verville Fellowship created the time and space to write the manuscript.

Early in our research, we were given an enormous (and enormously valuable) cache of documents on 1960s-era equal employment at the Marshall Space Flight Center by Molly Porter of the MSFC Historical Reference Collection. These documents were eye-opening in many ways. Our knowledge of the MSFC's struggle with integration was further bolstered by NASM curator and von Braun biographer Michael Neufeld. One day, in response to a question, Mike said, "You know, I have a bunch of documents on civil rights from von Braun's papers. You want 'em?" We thank Mike not only for these numerous and extraordinarily helpful documents, but more importantly for his ability to read Wernher von Braun's handwriting.

The first significant piece of internal NASA documentation we received on the subject of integration came from Neil deGrasse Tyson, director of the Hayden Planetarium, who gave us a copy of James Webb's 1963 letter to von Braun instructing him to hire more African Americans in Huntsville. This served as an important guidepost as we dove into the NASA archives. On our path through those archives, we were assisted further by Elaine E. Liston, archivist at the Kennedy Space Center; Gregg Buckingham, NASA history representative; Dr. Jennifer Ross-Nazzal, Johnson Space Center; Mark Scroggins, archivist at the Johnson Space Center; Lauren Meyers, University of Houston–Clear Lake archivist; and most of all by Elizabeth Suckow and Colin A. Fries in the History Office at NASA headquarters in Washington, DC.

One of the most valuable documents discovered in the course of our research was Peter Dodd's article on early '60s-era race relations in the space communities, which shed important light on everyday life. Over the years, we saw multiple allusions to this document but were told by many knowledgeable people that it had never been published and could not be obtained. We therefore want to express our deepest gratitude to NASM's 2012–13 Guggenheim Fellow, Teasel Muir-Harmony, for finding it and giving it to us.

Our chapter on George Carruthers would not be what it is without the help of NASM senior curator Dr. David DeVorkin, who remembered that a vast cache of interviews with Carruthers existed, and Joe Anderson, director, and Amanda Nelson, associate archivist, at the Niels Bohr Library and Archives, who helped us find them.

The perspective of the black press is only narrowly available through publicly accessible databases in the United States, so we are grateful to Jonathan Cohen, a student at McGill University in Montreal and a NASM intern in 2012, for giving us access to McGill's extensive online repository of back issues of black newspapers from across the U.S.

The project to find the pioneers and give their stories a place in history was conducted with the help of the public radio documentary producer Soundprint Media Center and the web design firms Swim Design and Bully Entertainment. In particular, Moira Rankin, Soundprint's senior producer, deserves mention for her generosity of spirit when the time came to turn this work into a book, and for her early and persistent insistence that George Carruthers be part of this story. We also want to thank Jared Weissbrot for all his help with this project. Bully Entertainment and Swim Design created the website Cape Cosmos (http://capecosmos.org), which makes the pioneers' stories (and those of the first women in the space program) accessible to a younger generation. At Bully, Carlson Bull and Janice Bernach should be singled out, and we want to thank Andrew Iskowitz and Laurie Swindull at Swim Design. We also thank Maria O'Meara and Jane A. Martin for their work gathering the images that appear in this book. Several people at Texas State Technical College in Waco, Texas, also helped in the early stages of the creation of this book. To George Wilhite, Lisa McNew, Dr. Gerry French, Debbie Moore, and Mark Long, a sincere thank you for all you did and all you do.

This work has been read, vetted, critiqued, and tweaked throughout the years that it was being produced. Drs. Susan and Ed Youngblood of Auburn University are especially connected to this evolution. From late nights when Steven was in graduate school during the 1990s to content reading of the near-final draft of *We Could Not Fail* in 2013, the importance of their advice and friendship through the years is beyond words. The members of NASM's Wednesday Writers Group were an invaluable source of assistance in the creation of the manuscript of this book and, in particular, Hunter Hollins was our most avid and helpful reader. We also want to thank Anke Ortlepp, NASM's 2011–2012 Verville Fellow, for her patience with the earliest drafts of this manuscript.

Finally, we express our thanks and gratitude to Robert P. Devens, Sarah B. Rosen, Molly Frisinger, Nancy Bryan, and everyone else at the University of Texas Press for their help in bringing *We Could Not Fail* to print.

Richard Paul and Steven Moss
2015

The people in the Space History Division of the National Air and Space Museum provided insight, perspective, encouragement, and camaraderie throughout my time there. My thanks go, in particular, to NASM museum educator Beth Wilson, who got my foot in the door at the Smithsonian and has remained a valued friend, advisor, and reader throughout the years that led to this publication. Also at NASM, I want to thank Trish Graboske for teaching me how to get a book published, Mychalene Giampaoli and LeRoy London for always keeping me mindful of the importance of preserving this history, and exhibit designer Heidi Eitel for always putting a smile on my face when I needed it.

A huge thanks goes to Dr. Roger Launius for finding Steven Moss's thesis, "NASA and Racial Equality in the South, 1961–1968," which had been sitting unnoticed on a shelf in the NASA History Office. Had I not asked and he not looked, I would not have met Steven, this book would not have been written, and the stories of these pioneers would not have been preserved.

I of course thank my darling wife, Renee, for her steadfast love and support through the highs and lows of this project, which were many. I also want to thank my mother, Abby Schine, which almost goes without saying.

Finally, I want to thank Dr. Margaret Weitekamp, NASM curator of the Social and Cultural Dimensions of Spaceflight. Three years ago, when this project had stalled, I wrote Margaret to ask whether the Smithsonian had any fellowship opportunities for people without advanced degrees. She told me about the Verville Fellowship, encouraged me to apply, championed my appointment against both precedent and opposition, and then, after I received the fellowship, coached, instructed, educated, and grounded me; edited and corrected this manuscript; and, when the initial draft was rejected, not only dragged me out of my disappointment but inspired me to restructure and then rewrite in ways that led to the volume you are holding.

Richard Paul
2015

Thanks for the liberty, Mr. Roberts. Thanks for everything.

When I started researching the connection between civil rights and NASA in 1995, few people believed it existed. But two people believed in me. To my Mom, June, and sister, Susan, thank you for your love, your support, and the constant reminder that "Dad would be proud."

Several of my professors at Texas Tech University contributed to the success of my work then and what it has become now. I am thankful to Dr. Ronald Rainger and Dr. Alwyn Barr for their guidance and advice while serving on my thesis committee. Thanks also to Drs. Otto M. Nelson and Brian L. Blakeley, who endorsed my research-grant applications and continue to influence my career. I owe special thanks to William Johnson, chief archivist, and June Payne at the Kennedy library; and to Linda Hanson and Bob Tissing at the Johnson library.

A lifetime's worth of friends deserves thanks and recognition for their support of me and this book throughout our years together. Fortunately you are too many to name, but a world without you and all that you represent is a less happy place.

In the 1990s I became a true believer in NASA's role in achieving racial equality and sought converts. Years later, Richard Paul became a believer. Thanks to our commitment to each other, the story, and our willingness to compromise, *We Could Not Fail* is the manifestation of that shared belief.

Twenty years ago a history professor said of this topic, "If there were anything there it would have been written by now." Now, it is written.

Steven Moss

2015

Index

Page numbers in *italics* indicate photographs and illustrations.